Applications of Digital Twins and Robotics in the Construction Sector

Editors

Dr. Syed Saad
Science, Engineering & Manufacturing Centre
Sandwell College
West Bromwich, United Kingdom

Kumeel Rasheed
Department of Civil Engineering
CECOS University of IT and Emerging Sciences
Pakistan

Syed Ammad
School of Engineering and Built Environment
Griffith University
Gold Coast, Australia

CRC Press
Taylor & Francis Group
Boca Raton London New York

CRC Press is an imprint of the
Taylor & Francis Group, an **informa** business
A SCIENCE PUBLISHERS BOOK

First edition published 2025
by CRC Press
2385 NW Executive Center Drive, Suite 320, Boca Raton FL 33431

and by CRC Press
4 Park Square, Milton Park, Abingdon, Oxon, OX14 4RN

CRC Press is an imprint of Taylor & Francis Group, LLC

Library of Congress Cataloging-in-Publication Data (applied for)

ISBN: 978-1-032-82863-3 (hbk)
ISBN: 978-1-032-85571-4 (pbk)
ISBN: 978-1-003-51874-7 (ebk)

DOI: 10.1201/9781003518747

Typeset in Palatino Linotype
by Prime Publishing Services

Preface

Humans and the construction sector have long been together in the race of progress and it is certain that one cannot win without the growth of the other. But with an advent of digital tech in the recent decade tremendous changes are producing long lasting impacts on the Human-Construction culture. A culture which finds a common ground in duo tech advancements, i.e., 'Robotics' controlled via 'Digital Twins (DTs)'. "Applications of Digital Twins and Robotics in the Construction Sector" centers itself around duo culture Human-Construction comparable to DTs-Robotics. This book also shares a unique perspective in helping the reader to develop an understanding of the United Nations Sustainable Development Goals (SDGs) connected to Humans, Robots and the future of our Construction Industry.

Digital twins and robotics are already known to cause rather massive improvements of productivity. Now, if you superimpose this synergistic equation on human workers in construction, thus we may merge them with BIM and other ergonomic protection devices to completely transform on site activities. This integrated approach holds a prospect of raising such essential characteristics as productivity, safety, and environmental Issues of construction to a new higher level. Noting the potential of change, the European Union has matched its Horizon Europe R&D plans for the aggregation of digital twins and robotics for fields including construction.

No overemphasis can be placed on the relevance of this book. We the construction professionals, researchers and policymakers, are standing in the precipice of the new age of construction—the age of digitalization and automation. Such a shift is not only a technological one; it modifies how construction is done and the expertise required. Just picture yourself standing in the middle of a city surrounded by the sounds and sights of the world constructing itself in front of your eyes and technology steadily carving out the features of the future for the construction industry. That once familiar day-to-day landscape is steadily being relinquished to the tender mercies of technology and we are being unleashed to new horizons. It is not just about machines replacing workers—this is about creating new professions–professions that will require new expertise, new ideas and, most importantly, a new comprehension of the digital world. Try to picture yourself as an innovator and not a follower of trends, but as a maker of them.

This book is not only a manual but also a companion as you read further opening the doors to the future. When it comes to digital twins and robotics in construction, by following this book, you will be ready to understand and to contribute to it. As you read through the pages of this book, let me remind you this book is action-oriented, a future blueprint, and a roadmap to prepare you for the next generation of construction leadership. Wishing you only the best in your reading journey forward from our dedicated editors and authors.

We extend our sincere gratitude to all the authors for their valuable expertise and insights contributed to this book. Their dedication and effort have enriched the content, making it a comprehensive and valuable resource. Each author's unique perspective has been crucial to the success of this collaborative endeavor. Moreover, Dr. Syed Saad is incredibly grateful to his parents, Syed Manzoor Ali Shah and Azra Yasmine, siblings and his wife, Dr. Ume Hani Naeem for their continuous support and endless encouragement. Without their considerate nature, bringing this book to fruition would not have been possible. Lastly, Kumeel Rasheed extends his heartfelt gratitude to his family - his father, Muhammad Sadiq, his mother, Rohina Sadiq, and his brother, Furqan Rasheed - for their generous support, understanding and sacrificing family time to allow him to dedicate himself to this book.

Acknowledgements

The editors would like to express their sincere gratitude to the researchers, collaborators and contributors whose dedication and expertise are reflected throughout this book. Their insights into digital twin innovation and ever-evolving capabilities of robotics have been priceless in shaping the direction of this work. This effort is as much theirs as ours, and their steadfast support has been crucial in tackling the hurdles and exploring future prospects within the construction sector.

We are deeply grateful to CRC Press for their constant cooperation and guidance during the editorial process. Their team, particularly Mr. Vijay Primlani (acquisitions editor), exhibited impressive persistence and professionalism, committed to achieving the highest standards of quality for this publication.

Finally, we extend our heartfelt gratitude to our families for their firm patience and understanding during the demanding process of compiling this book. Their motivation and belief in our efforts have been an ongoing source of strength.

Contents

Abbreviations

3D	Three-Dimensional
ADIT	Agile Data Integration Techniques
AI	Artificial Intelligence
AMRs	Autonomous Mobile Robots
APIs	Application Programming Interfaces
AR	Augmented Reality
AT	Assistive Technology
BAMB	Buildings as Material Banks
BIM	Building Information Modeling
BLS	Beginning Life Stage
BREEAM	Building Research Establishment Environmental Assessment Method
CAD	Computer-Aided Design
CCPA	California Consumer Privacy Act
CE	Circular Economy
CFC	Cross-Functional Collaboration
CNN	Convolutional Neural Networks
Cobots	Collaborating robots
CPS	Cyber Physical Systems
DDDM	Data-Driven Decision Making
DE	Data Exchange
DPOs	Data Protection Officers
DT	Digital Twin
DTA	Digital Twin Aggregate
DTI	Digital Twin Instance
DTP	Digital Twin Prototype
DTs	Digital Twins
ELS	End of Life
EPD	Environmental Product Declaration

ERP	Enterprise Resource Planning
EU	European Union
FESG	Financial, Environmental, Social, and Governance Dimensions
GDPR	General Data Protection Regulation
GHG	Greenhouse Gas
GIS	Geographic Information Systems
GIS	Global Information System
GNSS	Global Navigation Satellite System
GPS	Global Positioning System
HR	Human-Robot
HRC	Human-Robot collaboration
HRI	Human-Robot Interaction
HVAC	Heating, Ventilation, and Air Conditioning
ICT	Information and Communication Technology
IFC	Industry Foundation Classes
IoT	Internet of Things
IT	Information technology
KPIs	Key Performance Indicators
LCA	Life Cycle Assessment
LEED	Leadership in Energy and Environmental Design
LiDAR	Light Detection and Ranging
MEP	Mechanical, Electrical, Plumbing
ML	Machine Learning
MLS	Middle Life Stage
MTPE	Machine Translation Post-Editing
NDE	Non-Destructive Evaluation
PDTs	Personal Digital Twins
PID	Proportional-Integral-Derivative
PII	Personally Identifiable Information
PLC	Product Lifecycle
PLM	Product Lifecycle Management
PO	Physical Object
PPE	Personal Protective Equipment
PPM	Predictive and Prescriptive Maintenance
ROS	Robot Operating System
ROVs	Remotely Operated Vehicle
RQs	Research Questions
RUL	Remaining Useful Life

SAM	Semi-Automated Mason
SDGs	Sustainable Development Goals
SLAM	Simultaneous Localization and Mapping
SMEs	Small and Medium Enterprises
TBM	Tunnel Boring Machine
TENG	Triboelectric Nanogenerator
TRL	Technology Readiness Level
UAV	Unmanned Aerial Vehicles
UGV	Unmanned Ground Vehicle
UN	United Nations
VDC	Virtual Design and Construction
VO	Virtual Object
VR	Virtual Reality
VTBs	Virtual testbeds
XR	Extended Reality

Chapter 1

Fundamentals of Digital Twins in Construction

Muhammad Bilal,[1] *Abdul Haq,*[2] *Syed Saad,*[3,*]
Kumeel Rasheed[1] and *Syed Ammad*[4]

Overview

Digital Twin (DT) technology in construction involves creating a real-time virtual model of physical assets that continuously adapts to changes in the environment or operations, enabled by advancements in virtual modeling and data acquisition technology. DT technology improves efficiency, lifespan, and reduces operational costs through proactive and predictive maintenance cycles. It collects and analyzes data from embedded sensors to monitor and control physical objects, producing up-to-date information and enabling effective simulation. Despite its advantages, the construction industry has been slow to adopt DT technology due to its complexity and receptiveness of stakeholders (Madubuike et al., 2022). However, DT technology holds significant potential for improving design, construction, operation, and maintenance processes. Digital twins, which employ sensors, represent real-time data that construction managers, designers, or the customers use to follow projects in real-time. Construction teams utilize digital twins to monitor project progress, identify possible issues, and adjust their plans to ensure projects are

[1] Department of Civil Engineering, CECOS University of IT and Emerging Sciences, Pakistan.
[2] School of Civil Engineering, Southeast University, Jiulonghu Campus, China.
[3] Science, Engineering & Manufacturing Centre Sandwell College West Bromwich, United Kingdom.
[4] School of Engineering and Built Environment, Griffith University, Gold Coast, Australia.
* Corresponding author: syed.saad@sandwell.ac.uk

completed safely, on time, within budget, and to the desired quality standards. Additionally, digital twin solutions in the construction sector can aid in resource planning and logistics, safety monitoring, and tracking resources such as labor, materials, and equipment. In the real estate sector, digital twins can provide an accurate representation of physical assets, allowing agents or property owners to collect and analyze data on the asset's functionality and condition. The implementation of digital twin technology in the building and real estate industries leads to remarkable enhancements in cost, time, sustainability, and safety efficiency, making it an indispensable asset for the entire sector (Attaran and Celik, 2023).

In order to minimize resource waste during building construction, lean construction emphasizes achieving consistent and smooth production flows (Rasheed et al., 2023). Koskela's Transformation-Flow-Value theory enhances the traditional view of production in the construction industry by introducing the concepts of flow and value, providing a comprehensive understanding of production processes. In contrast to the conventional perspective, which emphasizes isolated activities within "black-boxes," lean construction focuses on process visibility and transparency. This method makes it possible to monitor and comprehend data produced by the digital twins more effectively, which improves workflow optimization. An efficient model of production planning and control can be built by combining the capabilities of the digital twin technology with the concepts of lean construction. Using the data from the digital twins, this approach optimizes construction workflows by locating bottlenecks, streamlining procedures, and making well-informed judgments. This integration leads to improved project outcomes, waste reduction, and enhanced efficiency in the construction sector (Sacks et al., 2020).

Sustainable growth can be achieved from the collection and application of data through the use of digital twins in the design, construction, operation, and maintenance of physical infrastructure assets (Broo and Schooling, 2023). One of the latest and most rapidly growing technologies, digital twins has enabled technological improvements that have helped the building industry. Building-related data, including progress, workers, materials, and equipment, can be gathered in real-time and visibly presented on the platform via digital twin technology. Through the digital twin platform, stakeholders access information and utilize data and models to make precise deployments for the upcoming project (Su et al., 2023). Digital twins offer numerous benefits in the construction sector.

- Digital twins reduce rework and related expenses by assisting in the early detection of possible design defects or construction problems.

- By simulating construction processes, digital twins make it possible to identify and reduce safety problems before they arise on the job site.
- Digital twins enable a comprehensive view of construction projects, promoting effective asset management throughout every stage of the project.
- By leveraging digital twins, it becomes possible to perform proactive maintenance and optimize operations through the tracking and evaluation of built asset performance.
- Through the use of digital twins, project stakeholders can collaborate in real-time, improving coordination and communication throughout the construction phase.
- Digital twins enhance project evaluation and forecasting while also helping to achieve sustainability and environmental goals.

Understanding Digital Twins

The term "Digital Twin" describes an integrated multiphysics, multiscale probabilistic, ultra-realistic simulation developed by NASA. This simulation can replicate physical models, historical data, real-time data, and other elements of a system or product to create an accurate digital counterpart (Ali et al., 2024). An autonomous system's behavior and its actual surroundings can interact with each other through a model known as digital twins. Through the use of physical, virtual, and interaction data, a digital twin is an actual mapping of every component in the product life cycle (Tao et al., 2019). A digital twin is a virtual representation of an object or system that covers its entire lifespan, is continuously updated with real-time data, and aids in decision-making through reasoning, simulation, and machine learning.

The concept of the digital twin was first utilized in the aerospace industry around 2002. Michael Grieves introduced the idea at the University of Michigan in 2003, defining it as a "digital twin or an equivalent to a physical product" (Opoku et al., 2021). A digital twin is a digital representation of a real-world item or system, essentially a digital replication of the physical world. The virtual version closely mirrors its real-world counterpart. Digital twins are being explored to enhance the performance of physical entities through computational techniques enabled by their virtual counterparts. Typically, digital twins consist of a physical entity, a virtual counterpart, and data connections between them (Jones et al., 2020). The modern concept of digital twins emerged from the integration of numerous technologies, allowing for real-time data collection, integration and analysis.

In the aerospace domain, a digital twin is like a super advanced simulation of an aircraft, using sensors and models to mimic its real-life counterpart. In the infrastructure domain, it is a digital version of buildings or processes, capturing every detail of the built environment in real-time (Ammad et al., 2021). For manufacturing, it is a digital copy of machines or parts, helping predict how it will perform in the production process. In healthcare, a digital twin acts as a virtual copy of physical objects, monitoring them in real-time without being physically present. In construction, it refers to a detailed digital replica of physical assets that is closely connected to the real-world objects it represents. Overall, the essential components of digital twin models are the physical elements, the connecting data, and the virtual representations. They are all about providing dynamic, real-time info for planning and control. Tools such as radio-frequency identification (RFID) and laser scanners help them stay connected to the physical world. Furthermore, by incorporating artificial intelligence (AI) or big data analytics, digital twins can become even more intelligent, enabling them to make autonomous decisions (Rasheed et al., 2024).

Digital Twin Technologies

The three main aspects of digital twins are data application, data modeling, and data collection (Lv and Xie, 2022). Digital Twin employs four technologies to collect and store real-time data, gather information for insightful analysis, and create a digital representation of a physical object as illustrated in Figure 1. These technologies include the Internet of Things, cloud computing, AI, and extended reality.

Internet of Things (IoT)

The Internet of Things (IoT) is a vast network of connected 'things'. This network includes relationships between things themselves, between people, and between people and things. IoT is the primary technology utilized in all digital twin applications. By 2027, over 90% of IoT platforms will support digital twinning. IoT employs sensors to collect data from real-world objects, creating an electronic replica of the physical item (Haris et al., 2024). This digital version can then be analyzed, modified, and optimized. IoT enables the continuous creation and updating of a real-time virtual representation of a physical object for digital twin applications. Therefore, IoT remains the cornerstone technology utilized in all Digital Twin applications (Attaran and Celik, 2023).

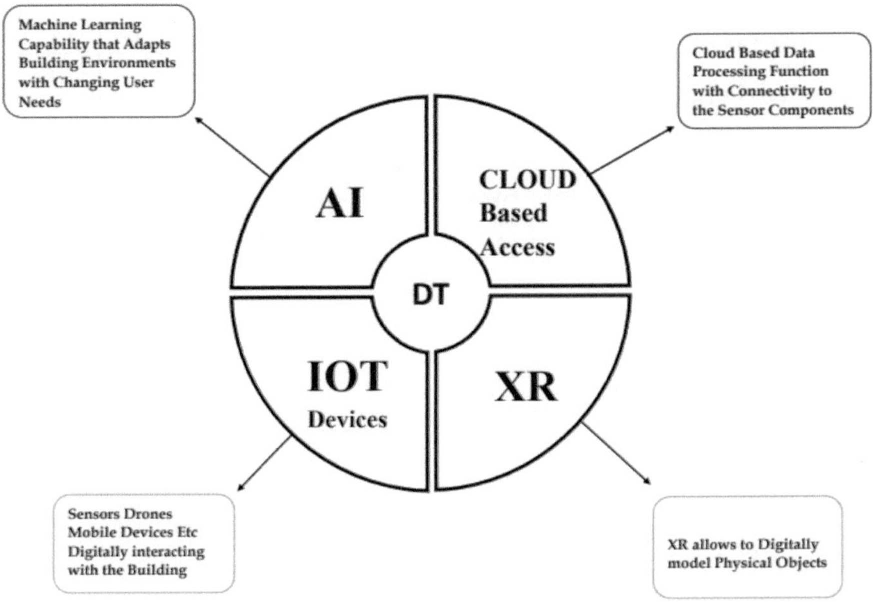

Figure 1: Digital Twin technology.

Cloud Computing

Cloud computing refers to hosting and delivering services over the Internet. This technology efficiently stores and retrieves data via the Internet. Digital Twins benefit from cloud computing by accessing data processing and cloud storage capabilities. Large data volumes can be stored by digital twins on the virtual cloud using cloud computing, which makes it simple to retrieve the necessary data from anywhere (Saad et al., 2024). Digital twins are able to efficiently minimize the processing time of complex systems and get over the challenges associated with storing huge volumes of data due to cloud computing (Attaran and Celik, 2023).

Artificial Intelligence (AI)

AI, or artificial intelligence, refers to the simulation of human intelligence by machines, especially computer systems. It encompasses various fields, such as robotics, language recognition, and image identification. AI can help digital twins by providing sophisticated analytical capabilities using methods like neural networks, machine learning, deep learning, and expert systems. It can autonomously analyze data, forecast outcomes, provide recommendations to prevent issues, and deliver insightful analyses (Attaran and Celik, 2023).

Extended Reality (XR)

Extended Reality (XR) refers to immersive technologies, including mixed reality, augmented reality, and virtual reality, that merge the real and virtual worlds to deliver enhanced experiences. These technologies enable users to interact with and visualize digital twins, which are virtual replicas of real objects or processes. In sectors like manufacturing, virtual reality and digital twin solutions combine to offer realistic problem-solving experiences. XR reduces costs and provides real-time direction while improving human-to-human contacts through virtual training and remote assistance (Attaran et al., 2024).

Characteristics of Digital Twin

Digital twins are virtual replicas of the physical geometry, behavior, and performance of assets, systems, or processes. They are continuously updated with sensor data, enabling precise monitoring and analysis. Two-way communication between the physical asset and its virtual counterpart simplifies performance control. Digital twins use AI, machine learning, and data analytics to offer insights, forecasts, and optimization suggestions for preventive maintenance and decision-making as shown in Figure 2. The key characteristics are:

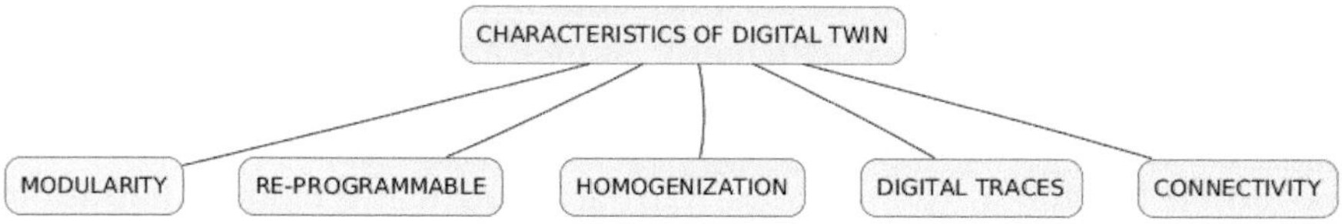

Figure 2: Characteristics of Digital Twin.

Connectivity

Connectivity is the foundation of a digital twin. It enables the connection between digital and physical elements. Sensors create this connectivity for the physical product by collecting, integrating, and transmitting data through various integration methods.

Homogenization

The two stages of homogenization include first aligning the data spatially and then making sure the data quality is compatible. Digital twins are a byproduct of data homogeneity as well as its enabler. It enables the abstraction of information from its physical embodiment.

Digital Traces

Digital traces left by digital twin technology can be useful in determining the root cause of machine faults. These traces act as a data trail that can be examined to identify the underlying source of the problems. Organizations can increase maintenance and operational efficiency by using this information to diagnose and solve issues more quickly (Hosamo et al., 2022).

Modularity

The term "modularity" describes how products and industrial modules are designed and customized. Modularity, when added to functional models, allows manufacturing to make modifications to models and machinery (Guo et al., 2019).

Re-programmable

Re-programmability is made possible by digital twins' use of sensors, AI, and predictive analytic techniques. Through ongoing data collection from its physical counterpart, the digital twin is able to modify its behavior in response to current circumstances. This makes it possible to adjust and optimize in real time in response to changing circumstances. The reprogrammable nature of digital twins enhances its ability to model, assess, and optimise the performance of the associated physical asset.

Evolution of Digital Twins in Construction

Industry 4.0, also known as the fourth industrial revolution, has brought digital technology, sensor systems, intelligent machinery, and smart materials to the construction sector. BIM has evolved into the primary tool for compiling digital project data. By utilizing modern technology to address the unique issues facing this industry, the term "Construction 4.0" refers to an effort to apply Industry 4.0 to the civil construction sector (Saad et al., 2024). The development of Building Information Modeling (BIM) based digital ecosystems has enabled the extraction of digital twins from real building environments. These digital twins can be integrated into construction management systems to assist in decision-making by keeping an eye on the tangible resources (people, materials, and equipment) at the construction site. To collect, store, and visualize data, appropriate auxiliary tools must be employed. This enables real-time monitoring of physical resources on construction sites and the display of relevant information (Ferreira et al., 2021). The rapid advancement of new technologies has led to an unprecedented industrial revolution in the modern era. One such technology, the digital twin, has garnered significant interest recently for its potential to transform the construction industry (Omrany et al., 2023). Construction has seen a revolution because of the development of digital twin technology, which offers a virtual version of real assets and operations. Beyond design and visualization, real-time monitoring, analysis, and simulation are also among its many uses (Zaland et al., 2024). As digital twin technology continues to progress and gains traction, it has the potential to completely transform the construction industry by enhancing project results and stimulating creativity.

This digitalization forms the foundation of smart infrastructure, represents the logical progression of infrastructure management systems, and enhances the overall design perspective (Waqas Khan et al., 2024). For example, smart infrastructure is the integration of physical infrastructure with sensor networks that provide real-time digital information on the condition of assets, and enhance decision-making and asset management (Saback et al., 2023). Over time, digital twins have undergone tremendous evolution that has increased their technological capabilities. They now collect data from a variety of sources, including BIM models, sensors, and IoT devices. This data integration provides a complete view of the project, facilitating performance analysis, predictive maintenance, and real-time monitoring of construction progress.

Advancements in AI and machine learning algorithms have significantly enhanced the capabilities of digital twins, enabling the implementation of predictive analytics and optimization algorithms (Rasheed et al., 2024). AI has the potential to assist in construction scheduling by predicting and managing time and cost contingencies for project completion, addressing risks related to quality, timeliness, and cost in every construction project (Khan et al., 2024). Among many other quantitative analytic techniques, these technologies can be used for natural language processing applications like word2vec, latent semantic analysis (LSA), latent dirichlet allocation (LDA), and clustering to save time and money on prefabricated construction projects. By leveraging AI technology, a pre-emptive contract risk evaluation can be conducted, minimizing claims and disputes among construction participants and providing stakeholders with contractual positions and rights based on factual contract data (Egwim et al., 2023). Digital twin technology was first applied in the construction industry for design and visualization needs. It enables stakeholders to explore different design options and identify potential issues before the construction begins. Additionally, it facilitates collaboration among contractors, engineers, and architects.

Digital Twin Technology Evolution Stages

Gartner's model for the advancement of digital twin technology is divided into three stages: replication of the real world, control, and optimization. A single product or system is frequently replicated in a virtual environment and subsequently optimized using simulation data obtained from numerous prior experiments conducted on the replicated model. However, since all phenomena and systems in the real world are organic and complex, relying solely on a digital twin for a single system may not produce the most optimal solution overall. Consequently, a five-phase model for the evolution of digital twin technology is proposed. The functional definition of each step is as follows.

1. **Mirroring:** Creating a digital twin of a physical object.
2. **Monitoring:** Observing and managing the tangible item through the examination of the digital replica.
3. **Modeling and Simulation:** Using the digital twin's simulation results to optimize the physical object.
4. **Federation:** Federated digital twin configuration, complicated physical object optimization, and interoperability between federated digital twins and complex physical things.
5. **Autonomous:** Automatically identifying issues in federated digital twins, resolving them, and optimizing tangible items in accordance with the federated digital twin solution (Jeong et al., 2022).

Types and Components of Digital Twins

There are three different kinds of digital twins: digital twin instance (DTI), digital twin aggregate (DTA), and digital twin prototype (DTP).

Digital Twin Instance (DTI)

The DTI specifies a particular physical product that has a corresponding digital twin that is tied to an individual digital twin throughout the duration of the physical object's existence. This type of digital twin can encompass various kinds of information, such as geometrical data, service information, part modifications, and more, depending on the specific use cases it is designed for (Grieves, 2016).

Digital Twin Aggregate (DTA)

This type of digital twin represents the entirety of the DTIs. Unlike the DTI, the DTA cannot be a distinct data structure; rather, it could be a computer architecture capable of querying all DTIs either on demand or in advance. A DTA provides an overarching view of all DTIs. Unlike a DTI, which is an independent set of data, a DTA might not be a standalone entity. It could be more like a smart tool that can access all the DTIs and ask them questions whenever needed, either on the spot or beforehand. For example, it might ask, "How often does component X usually fail?" On a proactive note, the DTA could keep an eye on sensor readings all the time and figure out patterns that could predict future failures (Grieves and Vickers, 2016).

Digital Twin Prototype (DTP)

This type of digital twin describes the prototypical physical artifact. It includes the data sets required to describe and create a physical copy of the virtual version, often known as a twin. These informative sets include the requirements, a fully annotated 3D model, a bill of materials (including material specifications), a bill of processes, a bill of services, and a bill of disposal, among other things (Grieves, 2016).

Components of Digital Twins

Although a digital twin is generally considered to be the interaction between a physical and virtual twin, all digital twins have five essential components: connectivity, data sources, virtual representation, data integration and analytics, and physical entity. A digital twin comprises the following elements.

Physical Entity

At the heart of the digital twin concept lie physical entities, which can be anything from a simple device to a complex system like a factory floor. These physical entities are real-world objects capable of independent action. The primary purpose of studying these entities is to monitor their functioning using various sensors. This aids in gathering data about their performance and current status, which is essential for understanding and improving them (Liu et al., 2023).

Virtual Model

The virtual model contains a digital representation of the physical object. This includes behavioral algorithms, geometry, material properties, and other relevant features. A virtual representation of a system or object that spans its entire lifecycle, continuously updated with real-time data, and aids decision-making through reasoning, simulation, and machine learning (Zaland et al., 2024).

User Interface

The interface comprises dashboards, visualization tools and control systems through which users interact with the digital twin. This enables them to monitor, analyze and manage the physical entity (Grieves, 2016).

Connectivity

Its operation involves communication networks and protocols that enable seamless data flow between physical objects and their digital twins. This ensures an instantaneous response and synchronizes the virtual model with the real object, providing an accurate and up-to-date representation. It serves as the backbone for integrating digital twin systems across various functionalities (Tancredi et al., 2022).

Benefits and Challenges of Digital Twins in Construction

Digital twin technology is still in its infancy within the construction industry but demonstrates promising benefits in various applications. One significant advantage of DT is evident during a facility's performance and operations, where they can deliver substantial returns on investment despite high initial costs. Digital twins offer numerous benefits, including effective data management, anomaly detection for maintenance, and enhanced access control. Reports by Deloitte and Johnson Controls emphasize additional benefits such as improved quality, reduced operational costs, and enhanced predictive capabilities for the future states of buildings. Overall, the application of digital twin technology in construction offers numerous advantages including automated progress monitoring, updated drawings/models, resource planning, safety monitoring, quality assessment, optimization of equipment usage, worker tracking, facilities management, decision-making, and sustainable development (Haris et al., 2024). Construction companies can utilize digital twins in various ways to address different aspects of their workflow. Developing custom solutions enables the creation of digital twins that are specifically tailored to meet the unique demands and specifications of a given team or project, thereby optimizing benefits and adding value to the company. The following are the primary benefits of using digital twin technology in the construction sector.

Resource Planning and Logistics

In construction, unnecessary movement wastes time and resources. Digital twin solutions improve resource management by providing real-time monitoring, thereby enhancing efficiency (Ali et al., 2024). For instance, integrating sensor data from equipment helps monitor asset usage and identify bottlenecks. These benefits of digital twins prevent resource over-allocation and improve time management in construction.

Building Performance Assessment

Digital twin solutions assess building performance by simulating their behavior in various scenarios. They achieve this by drawing data from multiple sources, such as sensors and models. For example, digital twins can evaluate a building's energy efficiency by simulating its energy consumption and generation, and experiment with different design approaches. It also assesses comfort and air quality by simulating airflow and humidity, testing various design and control strategies (Deng et al., 2021).

Construction Safety

The construction industry has a troubling safety record. Digital twin technology steps in by blending AI with video cameras, sensors, and mobile devices to create a robust safety framework for construction sites. With the real-time site reconstruction feature of digital twins, the construction process can be closely monitored as it happens, indicating any issues or deviations from the plan that could pose safety risks (Saad et al., 2024). This helps prevent the use of hazardous materials and activities in risky areas. Additionally, management can establish an early notification system to alert them when a worker is at risk, sending direct messages to the worker's wearable device. Furthermore, digital twins can spot unsafe behaviors and provide personalized virtual training for workers, cutting down on the need for expensive and time consuming physical training sessions and ultimately reducing the chances of accidents (Attaran and Celik, 2023).

Automate Project Control

Digital twins provide a valuable tool for monitoring construction progress by providing real-time updates on project status. This is achieved by integrating data from various sources such as sensors, drones, and scanners into a virtual model. AI-powered digital twin solutions can analyze this data to generate a detailed model of the actual construction progress, comparing it to the original plan on a daily or hourly basis. This tracking helps identify potential issues early, such as budget overruns or schedule delays, allowing the team to take prompt corrective actions to keep the project on track (Attaran and Celik, 2023).

Quality Control

In construction projects utilizing BIM, quality management entails linking 3D models with company-organized inspection checks. Effective site monitoring is crucial despite challenges in integrating site data, such as photos and scans, with BIM. A bridge construction case study with sensors demonstrates practical solutions to these monitoring and quality management challenges, highlighting the trend of integrating sensors into infrastructure for improved cyber-physical integration and future lifecycle insights (Boje et al., 2020).

Visual Communication

Visualizing complex data in multiple dimensions remains a challenge, requiring clarity and customization for user needs and collaborative meetings. Linear programming of construction schedules and vibrant color schemes aid in creating meaningful 4D models, especially for infrastructure maintenance. Integrating these models with site monitoring techniques enhances large-scale data management. Standardizing 4D visualization would need alignment with each collaborator's role and discipline, ensuring effective teamwork (Boje et al., 2020).

Sustainability

The use of Digital twins has a significant place in encouraging sustainable methods for construction work. They make it possible to perform in depth energy efficiency analyses and therefore contribute to the reduction of electrical power consumption. Digital twins reduce material consumption and waste by efficiently utilizing materials This technology not only guarantees that construction projects are environmentally acceptable but it also ensures that they are cost effective. In brief, the digital twins represent a huge leap toward sustainability in building and construction (Bortolini et al., 2022).

Real-Time Monitoring and Analysis

Real-time data can be captured, streamed to a digital platform, and shared with project stakeholders. Stakeholders can then monitor the project's current status in real time, perform analysis, and engage in closed feedback loops throughout the entire construction project lifecycle. This physical-digital connection is made possible by digital twins and is powered by the development of cyber physical systems (CPS), IoT, sensors, and AI. With these developments, digital twin technology provides a quick closed-feedback loop and insightful real-time data, making sure that everything that occurs on-site is recorded, monitored, and evaluated. For building projects, traceability is crucial, especially for complex projects. It can be difficult to trace and monitor people, products, equipment, materials, and information over the duration of a project. Construction stakeholders at different stages of the project can utilize real-time, up-to-date information obtained through tracking to inform and execute their responsibilities effectively. The integration of heterogeneous data sources by digital twin technology can enable real-time tracking and ensure traceability of the construction project and its various components (Ammar et al., 2022).

Challenges of Digital Twins in Construction

While digital twin technology offers numerous benefits, it currently faces challenges similar to those encountered by AI and IoT technologies. These challenges encompass data management, security, and standards as well as difficulties in implementation and transforming legacy systems. Challenges highlighted include the lack of a unified modeling method, the need to update outdated IT infrastructure, and difficulties related to connectivity, privacy, and security of sensitive data. Major challenges that could hinder the growth of the digital twin industry include high deployment costs, increasing demands for power and storage, challenges in integrating the technology with existing proprietary systems, and the complexity of the architecture involved. Implementing digital twin solutions is costly and demands significant investments in infrastructure development, ongoing maintenance, ensuring data quality control, implementing robust security solutions, and adopting advanced technological platforms. Additionally, it might be expensive to maintain the digital twin infrastructure, requiring a substantial operational investment. The adoption of digital twin technologies is likely to be impeded by their high initial costs and the complexity of their necessary infrastructure (Attaran and Celik, 2023).

Cost Challenges

Implementing digital twin technology necessitates significant initial investments in hardware and software, which can be prohibitive for small to medium-sized enterprises in the construction sector (Deng et al., 2016). Additionally, the return on investment (ROI) for digital twins can be challenging to quantify, potentially detering stakeholders from adopting this technology due to the associated costs (Matarneh et al., 2019).

Integration Challenges

The process of integrating digital twins with existing construction workflows and systems can be challenging. It involves not just technical updates but organizational and cultural changes as well. Resistance to change from employees in addition to stakeholders, together with a lack of uniformity on the way digital twin applications are implemented can make matters even more complex (Lu et al., 2020).

Data Management Challenges

Construction projects generate data from BIM, IoT devices, drones and sensors. It is very important to efficiently manage this data so that quality, reliability and relevance are maintained. Poorly-managed data can result in an inaccurate digital twin, undermining the system effectiveness (Bilal et al., 2016).

Technical Challenges

Developing dynamic digital twins and linking them with construction processes is highly technical due to their sophisticated, and ever-evolving nature. Reproducing real-world scenarios and processes with extreme precision requires advanced algorithms and significant computational power (Boje et al., 2020). Additionally, ensuring interoperability between various software and hardware systems used in construction is challenging due to the proprietary formats and standards for these systems (Khajavi et al., 2019).

Human Factor Challenges

For digital twins to succeed in construction the workforce must possess the necessary skills and capabilities. Rapid technological advancements require continuous training and education. Additionally, the shortage of digitally proficient professionals with expertise in both construction & related technologies presents a significant barrier (Ghaffarianhoseini et al., 2017).

Classification of the Barriers to DT Adoption in the Construction Industry

The barriers can be categorized into industrial barriers, stakeholder-oriented barriers, barriers related to construction enterprises, and technology-related barriers as shown in Figure 3.

Stakeholder-Oriented Barriers

In construction, the successful adoption and utilization of technologies like digital twins heavily rely on industry stakeholders embracing them. However, there is considerable confusion and disagreement

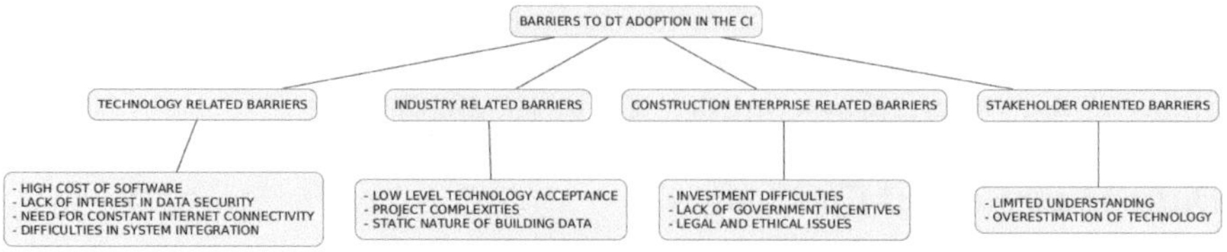

Figure 3: Barriers to DT adoption in the CI.

among professionals regarding the actual capabilities and benefits of digital twins. Many professionals in the construction industry do not have a complete understanding of the concept or the full potential of digital twins. This lack of clarity leads to misunderstandings and impedes progress. Compounding the issue, some confuse digital twins with, further complicating matters. There is no clear consensus on the comprehensive capabilities of digital twins in the industry, with some stakeholders only scratching the surface and not fully recognizing their potential for enhancing communication and efficiency. Additionally, there is an overestimation of advanced technologies like digital twins, BIM, and blockchain in solving industry issues. These misconceptions and lack of clarity hamper the ability to use digital technology effectively. The novelty of digital twins also makes it hard for people to see their value and how they fit in. Professional disconnect and fragmentation further complicate the adoption of emerging digital technologies like digital twins in construction. To make progress, we need better understanding and collaboration among stakeholders to fully leverage digital twins in construction projects (Opoku et al., 2023).

Construction Enterprise Barriers

Within the construction sector, individual organizations face challenges in adopting digital twins, partly due to the higher financial commitments involved. Cost considerations play a significant role in deciding whether to implement digital twins across different types of construction projects. Initial investments in digital twins can be substantial, adding to financial pressures for organizations. Investment difficulties, lack of government support, and legal/ethical concerns are identified barriers to smooth adoption. Economic challenges linked to digital twin utilization, as highlighted by Greif, Stein, and Flath, underscore the financial strains on construction enterprises (Greif et al., 2020). Some organizations may struggle to identify where to allocate investments effectively for maximum digital twin benefits. Government support for digital twin adoption remains limited, though initiatives like the UK's National Digital Twin Programme and Australia's efforts for centralized data sources show progress. Legal and ethical issues around data security and breaches further complicate digital twin adoption in construction (Rasheed et al., 2024). Addressing these barriers requires collaborative efforts from industry and government to ensure smoother integration of digital twins into construction operations (Opoku et al., 2023).

Industrial Barriers

In the construction industry, understanding digital twin technology is crucial for driving its adoption. However, a lack of knowledge and understanding of new technology can create resistance to change. Regrettably, the construction sector is frequently perceived as one of the least digitalized and slowest to adopt innovation. The industry's complexity, with its various trades and components, makes integrating digital technologies challenging. Supply chain fragmentation adds another layer of difficulty to adopting digital twins. Unlike BIM, which deals with static data, digital twin requires real-time information to reflect the current state of buildings or structures (Saad et al., 2023). Real-time dynamic data is essential for creating accurate digital twins and optimizing maintenance processes. Automation of the updating process is crucial for maintaining digital twins effectively. To fully leverage digital twins, the construction industry must adapt to using real-time data sensing and analysis effectively. This shift is necessary for the seamless implementation of digital twins in construction projects (Opoku et al., 2023).

Technological Barriers

Introducing digital twin technology in the construction industry encounters technological challenges such as system integration issues, data security concerns, and challenges related to scalability and internet connectivity. Integrating new technology like digital twins with existing outdated systems poses a major challenge. Data security is also a big concern, involving risks in data handling and protection of intellectual property (Saad et al., 2024). Scalability problems and the constant need for internet access further complicate digital twin's application in the construction industry. Hardware and software limitations delay effective data transfer between physical and digital realms, limiting digital twin's full potential. Even though digital twins can replicate physical entities in real-time, efficiently streaming data from the physical entity to the digital twin remains a challenge. Access to quality data for validating the accuracy and reliability of the digital twin is another hurdle. These challenges collectively impede the seamless integration and utilization of digital twins in the construction sector. Overcoming these barriers requires concerted efforts to address issues related to technology integration, data security, scalability, and data validation.

Application of Digital Twins in Project Planning and Design

The suggested real-time synchronization system for digital twins leverages their capability to integrate real-time data seamlessly into planning and execution processes for on-site assembly in prefabricated construction projects. Digital twins provide stakeholders with a comprehensive real-time view of resource status and construction progress, facilitating proactive decision-making. This system automates task generation, navigation, and schedule adjustments based on current data, thereby streamlining planning and execution processes. Inspired by functional tickets, digital twin real-time synchronization employs a ticket-based model to synchronize logistics and operations effectively. Moreover, its resilience to uncertainties ensures optimal resource allocation amidst dynamic conditions. Ultimately, digital twin synchronization optimizes planning by providing real-time visibility, facilitating task adjustments, and strengthening resilience in precast on-site assembly projects (Jiang et al., 2022).

Digital twin technology transforms construction by replicating physical assets virtually, aiding in planning and management. It aids in predicting project timelines and costs, as well as detecting potential issues beforehand, which helps in reducing delays. Smart modular systems integrating digital twin and robotics streamline on-site assembly. High-fidelity digital twin frameworks provide real-time resource and progress updates, improving planning and execution. Collaboration is enhanced through shared virtual environments, enabling stakeholders to visualize and decide together. Blockchain enabled digital twin platforms ensure real-time monitoring, secure communication, and data integrity, enhancing collaboration. In essence, digital twin adoption in construction boosts productivity, collaboration, and project outcomes (Omrany et al., 2023). Digital twins enable project managers to streamline their workflow, visualize projects, and make necessary adjustments before construction begins, thereby saving time and resources. The technology's capability to test and refine designs beforehand is a significant advantage, contributing to minimizing errors and delays. Designers and engineers can simulate real-world scenarios to better understand the functionality of buildings or infrastructure projects, thus improving overall design quality. For instance, they can model pedestrian and vehicular movement to enhance security and usability (Badshah et al., 2024). Digital twin technology also optimize the construction process itself by allowing managers to simulate sites and test various scheduling options. This helps identify the most efficient approach, reducing costs and enhancing productivity.

References

Ali, Z., Biglari, R., Denil, J., Mertens, J., Poursoltan, M. et al. (2024). From modeling and simulation to Digital Twin: Evolution or revolution? *SIMULATION*. https://doi.org/10.1177/00375497241234680.

Ali, Z., Saad, S., Rasheed, K. and Ammad, S. (2024). AI future perspectives and trends in construction. In *AI in Material Science* (pp. 239–261). CRC Press. https://doi.org/10.1201/9781003438489-11.

Ammad, S., Saad, S., Bashir, M. T., Qureshi, A. H., Altaf, M. et al. (2021). Construction accidents via integrating Building Information Modelling (BIM) with emerging digital technologies a review. *2021 3rd International Sustainability and Resilience Conference: Climate Change*. https://doi.org/10.1109/IEEECONF53624.2021.9668066.

Ammar, A., Nassereddine, H., Abdulbaky, N. and Aboukansour, A. (2022). *Digital Twins in the Construction Industry: A Perspective of Practitioners and Building Authority*, 8(June): 1–23. https://doi.org/10.3389/fbuil.2022.834671.

Attaran, M. and Celik, B. G. (2023). Digital Twin: Benefits, use cases, challenges, and opportunities. *Decision Analytics Journal*, 6(November 2022): 100165. https://doi.org/10.1016/j.dajour.2023.100165.

Attaran, S., Attaran, M. and Gokhan, B. (2024). Digital Twins and industrial Internet of Things: Uncovering operational. *Decision Analytics Journal*, 10(December 2023): 100398. https://doi.org/10.1016/j.dajour.2024.100398.

Badshah, I., Saad, S., Ammad, S., Rasheed, K., Khan, M. W. et al. (2024). Enhancing sustainable mobility: Lane change strategies for efficient traffic flow at stationary bottlenecks. *2024 ASU International Conference in Emerging Technologies for Sustainability and Intelligent Systems (ICETSIS)*, 1197–1202. https://doi.org/10.1109/ICETSIS61505.2024.10459583.

Bilal, M., Oyedele, L. O., Qadir, J., Munir, K., Ajayi, S. O. et al. (2016). Big Data in the construction industry: A review of present status, opportunities, and future trends. *Advanced Engineering Informatics*, 30(3): 500–521. https://doi.org/10.1016/j.aei.2016.07.001.

Boje, C., Guerriero, A., Kubicki, S. and Rezgui, Y. (2020). Automation in construction towards a semantic construction Digital Twin: Directions for future research. *Automation in Construction*, 114(January): 103179. https://doi.org/10.1016/j.autcon.2020.103179.

Bortolini, R., Rodrigues, R., Alavi, H., Vecchia, L. F. D., Forcada, N. et al. (2022). Digital Twins' applications for building energy efficiency: A review. *Energies*, 15(19): 7002. https://doi.org/10.3390/en15197002.

Broo, D. G. and Schooling, J. (2023). Digital twins in infrastructure: Definitions, current practices, challenges and strategies. *International Journal of Construction Management*, 23(7): 1254–1263. https://doi.org/10.1080/15623599.2021.1966980.

Deng, M., Menassa, C. C. and Kamat, V. R. (2021). From BIM to digital twins: A systematic review of the evolution of intelligent building representations in the AEC-FM industry. *Journal of Information Technology in Construction*, 26: 58–83. https://doi.org/10.36680/j.itcon.2021.005.

Deng, Y., Cheng, J. C. P. and Anumba, C. (2016). Mapping between BIM and 3D GIS in different levels of detail using schema mediation and instance comparison. *Automation in Construction*, 67: 1–21. https://doi.org/10.1016/j.autcon.2016.03.006.

Egwim, C. N., Alaka, H., Demir, E., Balogun, H., Olu-Ajayi, R. et al. (2023). Artificial Intelligence in the construction industry: A systematic review of the entire construction value chain lifecycle. *Energies*, 17(1): 182. https://doi.org/10.3390/en17010182.

Ferreira, E. de A. M., de Jesus, B. S. V. B., Araújo, C. S., Rodrigues, Y. C. M. and Costa, D. B. (2021). Digital twins to monitor physical resources at construction sites with web application. 050006. https://doi.org/10.1063/5.0070688.

Ghaffarianhoseini, A., Tookey, J., Ghaffarianhoseini, A., Naismith, N., Azhar, S. et al. (2017). Building Information Modelling (BIM) uptake: Clear benefits, understanding its implementation, risks and challenges. *Renewable and Sustainable Energy Reviews*, 75: 1046–1053. https://doi.org/10.1016/j.rser.2016.11.083.

Greif, T., Stein, N. and Flath, C. M. (2020). Peeking into the void: Digital Twins for construction site logistics. *Computers in Industry*, 121: 103264. https://doi.org/10.1016/j.compind.2020.103264.

Grieves, M. (2016). *Digital Twin: Mitigating Unpredictable, Undesirable Emergent Behavior in Complex Systems (Excerpt). August*. https://doi.org/10.13140/RG.2.2.26367.61609.

Grieves, M. and Vickers, J. (2016). Digital Twin: Mitigating unpredictable, undesirable emergent behavior in complex systems. *Transdisciplinary Perspectives on Complex Systems: New Findings and Approaches*, 85–113. https://doi.org/10.1007/978-3-319-38756-7_4.

Guo, J., Zhao, N., Sun, L. and Zhang, S. (2019). Modular based flexible digital twin for factory design. *Journal of Ambient Intelligence and Humanized Computing*, 10(3): 1189–1200. https://doi.org/10.1007/s12652-018-0953-6.

Haris, M., Saad, S., Ammad, S. and Rasheed, K. (2024). AI in fabrication and construction. In *AI in Material Science* (pp. 169–192). CRC Press. https://doi.org/10.1201/9781003438489-8.

Haris, M., Saad, S., Ammad, S., Rasheed, K., Aman, H. et al. (2024). Integrating sustainable practices for effective solid waste management in peshawar: Overcoming challenges and promoting environmental. *2024 ASU International Conference in Emerging Technologies for Sustainability and Intelligent Systems (ICETSIS)*, 1–7. https://doi.org/10.1109/ICETSIS61505.2024.10459445.

Hosamo, H. H., Nielsen, H. K., Alnmr, A. N., Svennevig, P. R., Svidt, K. et al. (2022). A review of the Digital Twin technology for fault detection in buildings. *Frontiers in Built Environment*, 8(November). https://doi.org/10.3389/fbuil.2022.1013196.

Jeong, D. Y., Baek, M. S., Lim, T. B., Kim, Y. W., Kim, S. H. et al. (2022). Digital Twin: Technology evolution stages and implementation layers with technology elements. *IEEE Access*, 10: 52609–52620. https://doi.org/10.1109/ACCESS.2022.3174220.

Jiang, Y., Li, M., Li, M., Liu, X., Zhong, R. Y. et al. (2022). Digital twin-enabled real-time synchronization for planning, scheduling, and execution in precast on-site assembly. *Automation in Construction*, 141: 104397. https://doi.org/10.1016/j.autcon.2022.104397.

Jones, D., Snider, C., Nassehi, A., Yon, J. and Hicks, B. (2020). Characterising the Digital Twin: A systematic literature review. *CIRP Journal of Manufacturing Science and Technology*, 29: 36–52. https://doi.org/10.1016/j.cirpj.2020.02.002.

Khajavi, S. H., Motlagh, N. H., Jaribion, A., Werner, L. C., Holmstrom, J. et al. (2019). Digital Twin: Vision, benefits, boundaries, and creation for buildings. *IEEE Access*, 7: 147406–147419. https://doi.org/10.1109/ACCESS.2019.2946515.

Khan, M. I., Khan, N., Saad, S. and Khan, M. W. (2024). Artificial Intelligence in construction materials. In *AI in Material Science* (pp. 47–64). CRC Press. https://doi.org/10.1201/9781003438489-3.

Liu, X., Jiang, D., Tao, B., Xiang, F., Jiang, G. et al. (2023). A systematic review of digital twin about physical entities, virtual models, twin data, and applications. *Advanced Engineering Informatics*, 55: 101876. https://doi.org/10.1016/j.aei.2023.101876.

Lu, Q., Parlikad, A. K., Woodall, P., Don Ranasinghe, G., Xie, X. et al. (2020). Developing a Digital Twin at building and city levels: Case study of West Cambridge Campus. *Journal of Management in Engineering*, 36(3). https://doi.org/10.1061/(asce)me.1943-5479.0000763.

Lv, Z. and Xie, S. (2022). Artificial intelligence in the digital twins: State of the art, challenges, and future research topics. *Digital Twin*, 1: 12. https://doi.org/10.12688/digitaltwin.17524.2.

Madubuike, O. C., Anumba, C. J. and Khallaf, R. (2022). A review of digital twin applications in construction. *Journal of Information Technology in Construction*, 27: 145–172. https://doi.org/10.36680/j.itcon.2022.008.

Matarneh, S. T., Danso-Amoako, M., Al-Bizri, S., Gaterell, M., Matarneh, R. et al. (2019). Building information modeling for facilities management: A literature review and future research directions. *Journal of Building Engineering*, 24(April): 100755. https://doi.org/10.1016/j.jobe.2019.100755.

Omrany, H., Al-Obaidi, K. M., Husain, A. and Ghaffarianhoseini, A. (2023). Digital Twins in the construction industry: A comprehensive review of current implementations, enabling technologies, and future directions. *Sustainability (Switzerland)*, 15(14): 0–26. https://doi.org/10.3390/su151410908.

Opoku, D. -G. J., Perera, S., Osei-Kyei, R. and Rashidi, M. (2021). Digital Twin application in the construction industry: A literature review. *Journal of Building Engineering*, 40: 102726. https://doi.org/10.1016/j.jobe.2021.102726.

Opoku, D. -G. J., Perera, S., Osei-Kyei, R., Rashidi, M., Bamdad, K. et al. (2023). Barriers to the adoption of Digital Twin in the construction industry: A literature review. *Informatics*, 10(1): 14. https://doi.org/10.3390/informatics10010014.

Rasheed, K., Saad, S., Ammad, S. and Bashir, M. T. (2024). Industry 4.0 and construction. In *AI in Material Science* (pp. 65–87). CRC Press. https://doi.org/10.1201/9781003438489-4.

Rasheed, K., Saad, S., Ammad, S. and Bilal, M. (2023). Integrating sustainability management and lean practices for enhanced supply chain performance: Exploring the role of process optimization in SMEs. *Engineering Proceedings*, 56(1). https://doi.org/10.3390/ASEC2023-16370.

Rasheed, K., Waqas, M., Saad, S., Zaland, A., Ali, Z. et al. (2024). Ethical and societal implications of AI in construction. In *AI in Material Science* (pp. 216–238). CRC Press. https://doi.org/10.1201/9781003438489-10.

Rasheed, K., Zaland, A., Saad, S., Ammad, S., Rostami, A. et al. (2024). History of AI. In *AI in Material Science* (pp. 15–46). CRC Press. https://doi.org/10.1201/9781003438489-2.

Saad, S., Ammad, S., Rasheed, K., Balbehaith, M., Qureshi, A. H. et al. (2023). Enhancing construction management in restoration through BIM-assisted photogrammetric 3D model accuracy assessment. *2023 International Conference on Sustaining Heritage: Innovative and Digital Approaches (ICSH)*, 100–103. https://doi.org/10.1109/ICSH57060.2023.10482845.

Saad, S., Haris, M., Ammad, S. and Rasheed, K. (2024). AI-assisted building design. In *AI in Material Science* (pp. 143–168). CRC Press. https://doi.org/10.1201/9781003438489-7.

Saad, S., Rasheed, K., Ammad, S., Ali, Z., Zaland, A. et al. (2024). The conclusive role of AI in Industry 4.0 construction materials. In *AI in Material Science* (pp. 262–270). CRC Press. https://doi.org/10.1201/9781003438489-12.

Saad, S., Rasheed, K., Ammad, S., Waqas Khan, M., Zaland, A. et al. (2024). Is AI the architect of tomorrow's materials in the age of Industry 4.0? In *AI in Material Science* (pp. 1–14). CRC Press. https://doi.org/10.1201/9781003438489-1.

Saad, S., Zaland, A., Ammad, S. and Rasheed, K. (2024). Robotics in construction. In *AI in Material Science* (pp. 115–142). CRC Press. https://doi.org/10.1201/9781003438489-6.

Saback, V., Popescu, C., Täljsten, B. and Blanksvärd, T. (2023). Analysis of Digital Twins in the construction industry: Current trends and applications. *Lecture Notes in Civil Engineering*, 350 LNCE: 1080–1088. https://doi.org/10.1007/978-3-031-32511-3_110.

Sacks, R., Brilakis, I., Pikas, E., Xie, H. S., Girolami, M. et al. (2020). Construction with Digital Twin information systems. *Data-Centric Engineering*, 1(6). https://doi.org/10.1017/dce.2020.16.

Su, S., Zhong, R. Y. and Jiang, Y. (2023). Digital Twin and its applications in the construction industry: A state-of-art systematic review. *Digital Twin*, 2: 15. https://doi.org/10.12688/digitaltwin.17664.2.

Tancredi, G. P., Vignali, G. and Bottani, E. (2022). Integration of Digital Twin, machine-learning and Industry 4.0 tools for anomaly detection: An application to a food plant. *Sensors*, 22(11). https://doi.org/10.3390/s22114143.

Tao, F., Sui, F., Liu, A., Qi, Q., Zhang, M. et al. (2019). Digital Twin-driven product design framework. *International Journal of Production Research*, 57(12): 3935–3953. https://doi.org/10.1080/00207543.2018.1443229.

Waqas Khan, M., Saad, S., Ammad, S., Rasheed, K., Jamal, Q. et al. (2024). Smart infrastructure and AI. In *AI in Material Science* (pp. 193–215). CRC Press. https://doi.org/10.1201/9781003438489-9.

Zaland, A., Saad, S., Ahmad, W., Ammad, S., Khan, M. W. et al . (2024). Enhancing mechanical properties of concrete by adding polypropylene fibers. *2024 ASU International Conference in Emerging Technologies for Sustainability and Intelligent Systems (ICETSIS)*, 771–778. https://doi.org/10.1109/ICETSIS61505.2024.10459663.

Zaland, A., Saad, S., Rasheed, K. and Ammad, S. (2024). Automation in construction industry. In *AI in Material Science* (pp. 88–114). CRC Press. https://doi.org/10.1201/9781003438489-5.

Chapter 2

Robotics in Construction
Transforming the Built Environment

Ali Khan,[1] *Sayed Muhammad Muiz Halimi,*[1] *Syed Saad,*[2,*]
Kumeel Rasheed[1] and *Syed Ammad*[3]

Introduction

Robotic technology has revolutionized the development sector, bringing a new level of precision, expertise, and safety to the built environment. This has fundamentally altered how building projects are planned and completed. The process of development has always been physically demanding, characterized by deliberate challenges and inevitable hazards. However, the emergence of robotic technology is transforming the sector, offering fresh perspectives on enduring problems and creating opportunities that designers, engineers, and planners never seemed to have imagined. The worldwide construction market is valued at about $10 trillion annually, yet it faces significant challenges such as a severe labor shortage, low adoption of new technologies, and weak to nonexistent productivity growth (Xiao et al., 2022). For example, at the start of 2017, the USA had approximately 200,000 unfilled construction jobs, and the economic output per worker had been declining for decades. Additionally, many large-scale construction projects exceed their budgets, generate significant material waste, and frequently experience accidents.

[1] Department of Civil Engineering, CECOS University of IT and Emerging Sciences, Pakistan.
[2] Science, Engineering & Manufacturing Centre Sandwell College West Bromwich, United Kingdom.
[3] School of Engineering and Built Environment, Griffith University, Gold Coast, Australia.
* Corresponding author: syed.saad@sandwell.ac.uk

These factors suggest that the construction industry could greatly benefit from increased automation, particularly through the use of robots. Consequently, numerous start-up companies are seeking to capitalize on this opportunity. While some intend to automate current building methods like bricklaying, others are investigating the potential of cutting-edge methods like 3D printing. Robotics technology is increasingly being utilized across all stages of development, from initial planning and design to on-site construction and maintenance. A wide range of robots exists, varying in size and design. Examples include unmanned aerial vehicles (UAVs) and autonomous drones used for space research and surveillance, as well as exoskeletons and automated arms for precise assembly and heavy lifting tasks. Despite being discussed among robotics professionals, there is little agreement on what defines a robot precisely. Science fiction often portrays robots as androids or humanoids with human-like traits. In industrial automation, robots are known for their monotony and high efficiency so how can we define a robot?

According to ISO 8373, the International Organization for Standardization provides a definition of a robot as "a controlled, partially autonomous mechanism that can move, manipulate, or position itself." According to the Robot Institute of America, a robot is defined as "a multifunctional manipulator that can be reprogrammed to perform a variety of tasks by moving materials, parts, tools, or specialized devices through variable programmed motions" (Kurfess, 2018). Merriam-Webster provides an additional definition that is "a machine that mimics human appearance and can carry out a variety of intricate human functions, like walking and talking" (Xiao et al., 2022).

In the construction sector, robotics technology is being driven by several causes, such as the pressing need to boost output, enhance quality assurance, and worker safety (Bloss, 2016). The need for quicker, more effective building that can keep up with the city's rapid growth grows along with its economy and people. The construction sector also has to deal with persistent issues such lack of labor, the need to retain qualified personnel, and reducing environmental effects. Robotics empowers individuals, lessens reliance on conventional construction, and improves resource efficiency to provide sophisticated answers to these problems. Furthermore, contrary to what was previously thought to be impractical or prohibitive, robots in the construction sector can exhibit originality and invention in construction, guaranteeing the success of complex procedures. Architects and designers can push the limits of environmental design and introduce a new age of architectural expression and sustainability by utilizing robots to carry out jobs like 3D printing, modular construction, and prototype design (Khosravani and Haghighi, 2022).

This chapter explores the development of robot technology in the construction sector, looking into its fundamentals, uses, advantages, and fixes issues arising from its application. To better understand how robots can revolutionize hillside environments, an analysis of research data, market trends, and future prospects is conducted. Better, stronger, and more inventive products are made by landscaping.

- The chapter explores the field of construction robots, encompassing its definition and application, tracking its development in the construction sector, and evaluating present market trends and future growth opportunities.
- Additionally, various construction robotics archetypes will be examined, including exoskeletons that enhance worker strength and endurance, aerial robots for site surveying and inspection, industrial robots for automating repetitive tasks, autonomous vehicles for heavy equipment and material handling, and soft robotics that offer adaptable and flexible solutions for construction challenges.
- Applications of robotics in the real world of construction, including automated masonry and bricklaying, the innovative use of 3D printing in construction projects, robotic methods for demolition and deconstruction work, the incorporation of drones for site monitoring and inspection, improvements in robotic welding and fabrication processes, and the use of autonomous machinery for earthmoving and excavation tasks.
- The reduction of labor-intensive tasks, the improvement of site automation and security, the mitigation of safety hazards and risks, and the increased efficiency and productivity achieved through automation are all important areas discussed in this chapter.
- Lastly, the chapter confers collaborative robotics which will also be examined, focusing on the dynamic relationship between humans and machines in the construction sector. Additionally, future directions and emerging technologies that will impact this field will be explored.

The Landscape of Construction Robotics

The goal of development robots is to transform the way we build the world by bridging the gap between innovation and the development industry's age. In essence, robotic technology is the configuration of automated equipment, robotic counters, and mechanical autonomy to carry out tasks that are traditionally completed by human laborers in designated areas. This includes a range of tasks from on-site planning and preparation to finishing the completed project. These mechanical frameworks can adapt accurately and efficiently to dynamic and complex situations, largely due to advancements in sensors, actuators, and analytics.

Apart from pure labor, mechanical autonomy also includes the notion of human-robot collaboration (HRC), wherein humans and robots cooperate to increase productivity and safety at development sites (Liu et al., 2021). This partnership encourages the integration of manual and mechanized tasks, leveraging the unique attributes of both people and robots. Additionally, mechanical technology development is in line with the newest advancements like Building Information Modelling (BIM) and Augmented Reality (AR), providing real-time data visualization and forms for decision-making (Sadhu et al., 2023). This allows the development team to improve processes and minimize method errors.

Robotics is progressing beyond traditional research to support modern methodologies such as isolated research and additive manufacturing (3D printing). These progressions make it possible to enhance growth forms, cut waste, and provide more notable adaptability to satisfy the demands of modern development. In order to facilitate innovative techniques like additive manufacturing and modular construction, robot construction is expanding beyond traditional construction. Utilizing these technologies can lead to more effective development strategies, reduced waste, and greater flexibility to meet the demands of modern construction. Development robots have the power to revolutionize the sector and enhance natural supportability, productivity, and security. The way we conceptualize, plan, and construct the physical environment is being drastically altered by construction robots, which integrate automation, teams, and creativity.

Evolution of Robotics in Construction

From its conception to the current transformative role within the industry, robotics in construction has been a spectacular journey that holds significant implications for the future. Construction robots started out with simple automated equipment for repetitive jobs like mixing concrete and installing bricks. These early systems showed that automation has the ability to supplement human labor and increase production, despite certain limits. Technological developments quickly brought sensors and actuators together, improving robot perception and interaction with their surroundings (Saad et al., 2020). This greater autonomy opened the door for mobile robots that can now navigate building sites on their own and carry out a variety of duties, from material transportation to site inspection. These advancements decreased human exposure to dangerous situations, which increased safety in addition to increasing efficiency (Rasheed et al., 2021).

The widespread adoption of modern construction automation, which incorporates robotics, did not happen until the introduction of the first industrial robots in the 1950s and their utilization by the automobile industry in the 1960s. Factory automation was extensively employed in the industrial domain, whereas construction robots evolved during the 1960s and 1970s. In the 1970s and 1980s, Japan addressed a shortage of construction workers due to an aging population and disinterest among younger people by adopting construction automation and robotics (Toivonen, 2012). Specialized robots and remote-controlled machinery were developed for various building tasks, including excavation, material handling, concrete placing and finishing, fireproofing, earthworks, and rebar placement.

The creation of efficient robotic arms with complex manipulators and end-effectors is a significant milestone in the field of construction robotics. These tools are modified for operational purposes because of their outstanding accuracy and agility in activities like welding, assembling, and demolition (Rasheed et al., 2024). Another notable advancement that has surfaced is collaborative robotics, commonly referred to as cobots, which facilitate safe interaction between humans and robots at construction sites. Cobots possess the capacity to support human workers by offering assistance in tasks that require substantial physical effort or demand accuracy, while simultaneously guaranteeing a secure working environment as shown in Figure 1 (Schöberl et al., 2023).

Artificial intelligence (AI) has enabled robots to learn, adapt, and make decisions on their own, which has advanced construction robotics. AI-powered robots may review building designs, accelerate processes, and even foresee potential threats, enhancing project management and lowering errors (Pan and Zhang, 2021). Furthermore, by quickly

Figure 1: Robots working alongside a construction worker.

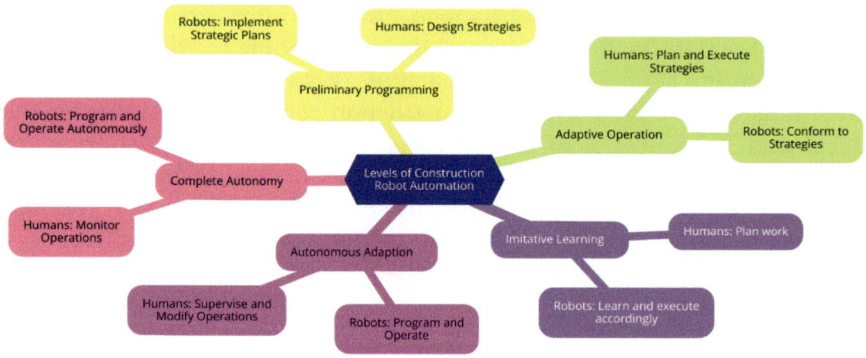

Figure 2: Levels of construction robot automation.

understanding and responding to human commands and gestures, AI-driven robotic systems improve human-robot interaction (Haris et al., 2024). A significant accomplishment in the field of additive manufacture, also known as 3D printing, is robotic construction. This method makes it possible to directly manufacture complex architectural components and structures on-site, which reduces material waste and eliminates the need for conventional assembly techniques (Haris et al., 2024). Since 3D printing offers unparalleled personalization and design freedom, it is revolutionizing sustainable construction processes and architectural innovation.

Future advancements in creating robotics seem to be happening quickly. Progress in robotics, artificial intelligence, and materials science will surely lead to the development of intelligent, autonomous robotic systems that can perform tasks with higher precision and efficiency. Block chain and augmented reality (AR) are also expected to enhance the capabilities of construction robots (Sadhu et al., 2023). Data visualization, supply chain management, and on-site construction team cooperation will all benefit from this (Naeem et al., 2022). Throughout its innovative and progressive history, robotics in construction has expanded its boundaries and redefined what is possible in the built environment with each new milestone (Saad et al., 2022). This trajectory from simple automation to intelligent robotics, represents the relentless pursuit of efficiency in the building sector, sustainability, and safety as specified in Figure 2. Additionally, it predicts a future where machines will play a critical role in revolutionizing the physical infrastructure.

Market Trends and Growth Prospects

Robots are bringing a revolution in the construction sector. The market for robots is in early stages but it is expected to develop significantly.

Businesses in industrialized nations like North America and Europe are investing heavily in robotic technology as it will increase production and worker safety (Leigh and Kraft, 2018). Several key trends are causing this spike as shown in Figure 3, including the following:

- **Labor Shortages:** Businesses are looking for automation solutions as a result of a global lack of qualified construction workers. This gap can be filled by robots, who can do laborious and physically demanding tasks while increasing overall production.
- **Safety Concerns:** There is no doubt that the construction sector carries a significant degree of danger. Robots can do dangerous tasks thus lowering the risk of accidents and worker fatalities.
- **Technological Developments:** Due to the developments in AI, robotics, and sensor technology, the robots are now equipped to perform challenging construction tasks.
- **Growth:** More than 15% annual growth is expected till 2029.

A key characteristic is the growing diversity of robotic applications within the construction robotics sector. In the past, the main uses for these applications were for concrete pouring and bricklaying. Robots can execute a variety of tasks. This diversification includes a range of tasks, including project management, material processing, and site surveys. As a result, robotics is providing creative ways to make operations more efficient and reduce project delays. This phenomenon is driving the progress of customized robotic systems designed specifically for different building jobs, addressing the diverse needs of construction organizations.

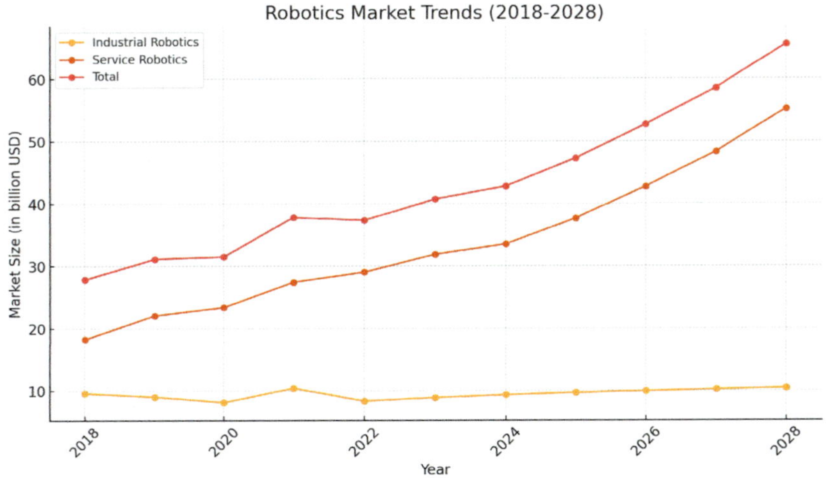

Figure 3: Expected robotics Industry's revenue by 2028 (in billion USD).

AI and machine learning enable robots to learn from experience, adapt to changing conditions, and make autonomous decisions (Rasheed et al., 2024). This enhancement of capabilities improves the variety and effectiveness of robots, making construction activities more efficient and responsive. AI-powered robots have the ability to examine data collected from sensors and various sources in order to improve workflows, predict potential dangers, and increase the overall results of a project (Saad et al., 2024). As AI continues to progress, it is expected that construction robots will become extremely proficient machines that can navigate complex construction sites with ease. Along with technology improvements, regulatory actions and industry standards aimed at improving sustainability and safety are driving the growth of the construction robot's market. Globally, governments and regulatory bodies are realizing more and more how robotics can enhance worker safety and lessen the environmental impact of construction projects. Consequently, legislation, financial incentives, and regulatory measures are being implemented to promote the integration of robotic technologies into construction processes. Furthermore, rules and best practices for the safe and efficient use of robots on construction sites are being produced by industry associations and standards organizations in collaboration, which is propelling the market's growth and uptake.

The future outlook for construction robotics appears promising. The industry is predicted to grow due to favorable legal frameworks, ongoing innovation, and a rising need for automation. According to estimates, robotics will become considerably more common in the residential, commercial, and infrastructure development sectors of the building industry (Piyush et al., 2021). It is anticipated that smaller companies and subcontractors would employ automation to stay competitive as robot technology advances and becomes more widely available. Ultimately, construction robotics have the potential to fundamentally transform the construction sector by offering previously unheard-of chances to boost sustainability, productivity, and safety.

Industrial demand, technological innovation, and regulatory backing have come together to form the market trends and growth potential of construction robotics (Waqas Khan et al., 2024). Construction companies can enhance efficiency, safety, and sustainability in construction activities by adopting automation, expanding applications, and utilizing modern technologies.

Diverse Archetypes of Construction Robotics

There is currently a significant increase in the use of robots in the construction sector, with a wide range of machines being developed to handle different duties across the entire building process. The robots can be classified into four primary archetypes, each possessing unique strengths and specialized applications:

Industrial Manipulator Arms

Frequently called robotic arms as shown in Figure 4, these adaptable robots imitate the functional anatomy of the human arm, featuring several joints and a gripper located at the extremity. With their advanced precision and programmability, these devices perform exceptionally well in regulated conditions such as prefabrication facilities (Saad et al., 2024). Typical uses involve the processes of welding, assembling pre-made parts, laying bricks, and applying sealants. The progress in sensor technology allows them to manipulate fragile materials and adjust to small deviations.

Figure 4: Industrial manipulator arms in construction.

Mobile Robots

Whether wheeled or tracked, these autonomous or remotely operated vehicles navigate construction sites with enhanced mobility and accessibility (Badshah et al., 2024). They are particularly effective for performing repetitive tasks in ever-changing environments. Examples consist of:

- **Delivery Robots:** These industrious machines have the ability to transport materials such as bricks, rebar, and drywall autonomously on the construction site, hence decreasing the need for manual labor and enhancing efficiency. With the aid of sensors and obstacle avoidance algorithms, they are able to move securely around workers and other equipment.

- **Demolition Robots:** These strong machines, frequently operated from a distance to ensure the safety of the operator, are fitted with hydraulic hammers or shears to dismantle concrete, walls, and other constructions as shown in Figure 4. They can enter restricted areas and handle dangerous demolition assignments.
- **Inspection Robots:** These robots are equipped with cameras, LiDAR (Light Detection and Ranging), and other sensors. They move through rough terrains and complex structures to gather data for inspections (Lattanzi and Miller, 2017). They have the ability to detect possible safety risks, pinpoint fractures or imperfections, and conduct visual evaluations of inaccessible regions. In Figure 5, the smaller robot, the Clearpath Robotics Jackal UGV (Unmanned Ground Vehicle), which is frequently utilized for research and development. In contrast the larger Clearpath Robotics Husky UGV is employed in a variety of outdoor applications, including research, agriculture, and inspection.

Figure 5: Delivery robots in construction.

Aerial Robots

Aerial robots are flying robots that offer a wide range of applications, from surveying and monitoring to logistics and safety inspections as illustrated in Figure 6. They are also called drones. Drones are extremely valuable and efficiently cut down time. They can improve productivity, reduce costs, and enhance safety on job sites. They are particularly valuable for:

- **Site Surveying and Mapping:** Drones fitted with high-resolution cameras and LiDAR can swiftly gather aerial data of the construction site, aiding in site surveying and mapping (Choi et al., 2023).

Figure 6: Aerial robot in construction.

- **Structural Inspections:** Drones can visually inspect hard-to-reach areas like roofs and facades to detect damage or potential safety hazards. Additionally, thermal imaging drones can identify heat loss in structures (Choi et al., 2023).

Real-World Applications of Robotics in Construction

The varied capabilities of construction robots are yielding significant benefits in real-world projects. Here's a closer look at some of the most impactful applications:

Automated Bricklaying and Masonry

The bricklaying industry, traditionally dependent on manual labor is undergoing a technological revolution with the advent of robots. These automated bricklaying robots can significantly boost production and reduce reliance on skilled workers due to their precision and accuracy. Utilizing computer-aided design (CAD) data, these robots can place bricks accurately and consistently, thereby minimizing the risk of human error (Aguair and Behdinan, 2015). Certain robots utilize a robotic arm equipped with a specialized gripper to handle the task of picking up and positioning bricks as shown in Figure 7. In contrast, other robots utilize a gantry system that incorporates a moving head to apply mortar and position the bricks. The method is especially beneficial for extensive projects with repetitive layouts or elaborate brickwork patterns.

3D Printing in Construction

Additive manufacturing, often known as 3D printing, is the latest technology used in the construction industry. Using a computer-controlled

Figure 7: Bricklaying cobot.

printer, it builds structures by layering materials like concrete (Hossain et al., 2020). It enables the formation of complex and natural forms that would be difficult or unattainable with conventional techniques. This enables the exploration of novel architectural opportunities and simplifies the creation of buildings that are both lightweight and robust. 3D printing reduces construction waste by utilizing only the required materials for the design, unlike traditional methods. 3D printers can be installed directly at construction sites, hence minimizing the requirement for transporting pre-manufactured elements. This can be especially advantageous in remote locations or for projects with complex logistical requirements. Although 3D construction printing is currently in its infancy, the accomplishments of many projects serve as evidence of its capacity to profoundly transform the building sector (Ammad et al., 2021). With its ability to print concrete walls for single-family houses and fabricate intricate bridge components, 3D printing is set to increasingly influence the construction industry.

Robotic Demolition and Deconstruction

Due to the hazardous materials and challenging conditions associated with it, demolition poses significant risks. Robotic demolition devices are revolutionizing this process by enhancing safety, efficiency, and precision. Equipped with tools such as hydraulic hammers and shears, these machines are often operated remotely and can dismantle steel, concrete, and other materials with reduced dust and a lower risk of accidents compared to traditional methods as shown in Figure 1. Robotic systems can also be configured to demolish certain areas, making it easier to rescue materials that can be used again for environmentally friendly building techniques.

Site Inspection and Monitoring with Drones

Drones fitted with various sensors and high-resolution cameras are invaluable for inspecting and monitoring construction sites. These UAVs can swiftly capture aerial images of the entire site, providing stakeholders and project managers with valuable insights (Lattanzi and Miller, 2017), as illustrated in Figure 6. Here are some ways drones are utilized:

- **3D Modeling and Progress Tracking:** High-precision 3D representations of the construction site can be made using data from drones. Project managers can utilize the collected data within the BIM system once it has been collected. In conjunction with BIM, these models enable project managers to systematically monitor project progress, detect deviations from the original plan, and promptly implement necessary adjustments (Saad et al., 2023). Project teams can make proactive decisions and guarantee that project objectives are regularly accomplished in real time with this all-inclusive approach, which gives them a dynamic framework for ongoing evaluation and adaptation.
- **Safety Inspections:** Drones can be employed to conduct visual inspections of hard-to-reach areas such as building facades or rooftops, to identify possible safety issues and reduce the risk to human inspectors. Additionally, when equipped with thermal imaging cameras, these gadgets can identify heat loss in buildings, simplifying energy efficiency studies (Choi et al., 2023).

Robotic Welding and Fabrication

Robotic welding and fabrication are becoming commonplace in construction procedures, fundamentally transforming the methods of creating and assembling structures. Robotic technology provides several advantages over manual methods including precision, efficiency, and high-quality welding (Kerber et al., 2018). Construction organizations can get exact control over welding parameters, such as arc duration, travel speed, and deposition rate, by utilizing robotic welding systems that are equipped with powerful sensors, controllers, and programming algorithms. Furthermore, robotic welding systems can weld complicated pieces faster and more accurately than human welders in difficult situations like tight spaces or hazardous environments. This lowers the possibility of mistakes, rework, and accidents. Robotic welding and fabrication greatly enhance efficiency and production on building sites. Robotic welding systems can be configured to perform tasks such as tack welding, multi-pass welding, and other welding processes automatically. This enhances production efficiency and reduces the need for human intervention (Efimov et al., 2019).

Robotic fabrication and welding further improve site safety by reducing the amount of time and effort that construction workers are required to physically exert themselves, breath in dangerous welding fumes, and endure severe heat. Consequently, there will be a decrease in injuries and health problems caused by welding. Modern safety features are integrated into robotic welding systems. The features included sensors that detect impending collisions, emergency stop buttons, and protective barriers (Wang et al., 2020). This safeguards and ensures that the machinery can operate in proximity to people without endangering them. More contemporary welding processes and materials can be used with robotic welding and fabrication, which would be impossible or extremely impractical to accomplish with traditional human methods. Using robotic welding technology, building companies can increase production and efficiency, find new structural designs and construction channels, enhance site safety, and obtain precise control over welding operations. The future of building and assembly will likely change significantly, as robotics technology evolves.

Excavation and Earthmoving with Autonomous Machinery

Excavation and earthmoving are essential construction operations required for the development of structures, services, and landscapes. The introduction of completely self-governing machinery has brought about a fundamental change in the construction industry due to progress in robotics. Implementing this technology on building sites enhances productivity, accuracy, and security. Digging and excavating machines are examples of autonomous machinery (Frank et al., 2019). These robots achieve autonomy through the integration of high-tech sensors, global positioning systems, and onboard computers. Comparing autonomous systems to traditional, human-operated ones reveals numerous advantages (Rasheed et al., 2023). They are cheaper, safer, and more effective. Autonomous bulldozers and excavators have the significant advantage of being able to operate continuously. These self-operating machines can reliably and accurately perform repetitive tasks such as leveling surfaces, transporting items, and digging ditches. This capability significantly reduces the need for human labor, minimizes errors, and decreases rework. Additionally, autonomous machinery enhances precision and accuracy in excavation and earthmoving operations, resulting in superior quality outcomes and reduced material wastage. Modern safety measures on autonomous machinery include collision avoidance systems, emergency stop mechanisms, and obstacle detection sensors (Khan et al., 2024). These elements are customary and aim to augment safety by offering an extra level of safety when operating in dynamic and unforeseen conditions.

Autonomous equipment that is environmentally friendly consumes less fuel, emits fewer emissions, and generates less noise compared to diesel-powered equipment. By maximizing equipment utilization, engine efficiency and route planning autonomous machinery can enhance sustainability and lessen environmental impact on building projects. This strategy reduces carbon emissions and fuel usage. Autonomous gear has the ability to operate with less noise compared to conventional equipment in delicate or urban environments, thereby safeguarding the well-being of local residents and preserving wildlife habitats. Due to advanced sensors, GPS systems, and onboard computers (Rasheed et al., 2021), these autonomous machines are able to perform excavation and earthmoving activities with exceptional efficiency, precision, and safety. As a result, they speed up project timelines, improve the quality of outputs, and reduce environmental effects. As robotics technology advances, autonomous machinery is becoming increasingly vital in the construction industry (Frank et al., 2019).

Robotics in Construction: Enhancing Site Automation and Safety

Robotics' recent introduction into construction procedures has completely changed the sector and improved efficiency and safety on construction sites. As we enter an era of digital construction, robotic technologies have become essential for automating operations that previously required significant human effort and posed safety risks. Robotics introduces a new paradigm in the approach, management, and execution of construction tasks, encompassing everything from improved surveying and monitoring to precision welding and material handling.

Table 1 shows the various functions that robotics performs in contemporary construction, highlighting the ways in which they enhance worker safety, boost production, and produce better project results. Detailed discussions on each robotic technology will follow, offering insights into how they specifically enhance automation and safety in the construction environment.

Increased Efficiency and Productivity

The integration of robotics into construction processes has fundamentally transformed the industry, notably enhancing efficiency and productivity. Table 2 summarizes the operational benefits and impacts of robotic bricklaying systems.

Table 1: Impact of robotics on construction site operations.

Robotics Application	Function and Benefit	Impact on Construction
Robotic Welding	Automates welding tasks with high precision and efficiency	Reduces labor needs, enhances structural integrity and speed of assembly
Autonomous Excavators	Use advanced sensors for precise and efficient excavation	Increase safety, reduce human error, and speed up ground preparation
UAVs for Surveying	Conduct aerial surveys with high-resolution cameras and LiDAR	Improve accuracy of site assessments, help in project planning and monitoring
Robotic Bricklaying	Automates the laying of bricks, operates continuously	Increases productivity, maintains consistent quality and reduces physical strain on workers
Modular Construction Robots	Assemble prefabricated components on-site with accuracy	Streamline construction process, reduce waste, and shorten project timelines
Drones for Monitoring	Monitor construction sites for compliance and progress	Enhance project management, ensure adherence to safety and design standards
AMRs (Autonomous Mobile Robots)	Transport materials and tools across sites efficiently	Improve logistics, reduce manual labor, and minimize the risk of injuries

Table 2: Overview of robotic bricklaying systems in construction.

Feature	Description	Benefits
Operation	Automated process of laying bricks Robots use advanced algorithms to position each brick precisely and apply mortar	Increases precision and consistency, reducing material waste
Productivity	Capable of operating continuously without fatigue Can lay bricks faster than manual methods	Significantly boosts the speed of construction, leading to shorter project timelines
Safety	Reduces physical strain on workers by handling repetitive, labor-intensive tasks	Lowers risk of injuries associated with manual bricklaying
Efficiency	Optimizes the use of materials and time. Robots can adjust techniques based on environmental conditions	Enhances overall construction efficiency, reduces downtime and operational costs
Integration	Easily integrates into existing construction workflows Can work alongside human workers or independently	Flexible deployment, supports a collaborative work environment with minimal disruption
Quality Control	Maintains high standards of construction quality, equipped with sensors to monitor and adjust the laying process	Ensures structural integrity and adherence to design specifications

Embracing robotic technologies not only reduces project timelines and labor costs but also improves safety and quality across construction projects (Saad et al., 2024). As the industry advances, the integration of diverse robotic systems—including both terrestrial and aerial technologies—is proving essential for the sustainability and competitiveness of the construction sector.

Reduction of Labor-Intensive Tasks

The incorporation of robotics into the construction sector signifies a significant transformation in the approach to labor-intensive jobs, providing a wide range of advantages that span from improved efficiency to strengthened safety measures. Historically building tasks such as bricklaying, welding, and excavation have predominantly depended on physical labor; this method is naturally prone to mistakes, time limitations, and potential dangers to safety. Nevertheless, the introduction of robotic technology has accelerated a change in these areas, fundamentally reshaping the standards of construction methods.

The labor-intensive tasks like trenching and foundation excavation are evolving with the advent of robotic excavation equipment. Equipped with advanced sensors and navigation systems, robotic excavators demonstrate exceptional accuracy and productivity. The inspection procedures which were heavily dependent on labor and were very time-consuming, have been simplified with the use of drones for monitoring and aerial inspections (Zaland et al., 2024). In order to provide stakeholders and project managers with useful information, drones equipped with LiDAR sensors and high-resolution cameras take precise pictures of construction sites. Construction organizations can optimize project timetables and overall efficiency by using drone technology to quickly identify and handle possible difficulties. With the advancement of technology labor intensive tasks have been significantly reduced.

Precision and Accuracy in Construction Processes

Site mapping and surveying are two areas where robotics has significantly improved accuracy and precision (Boya et al., 2023). When it comes to assembling building components, robots are essential for increasing efficiency and precision. Inefficient and error-prone manual assembly techniques are common in traditional building systems. On the other hand, robotic assembly technologies have completely altered the construction industry. Wall panels, floor systems, and roof trusses are just some of the prefabricated components that these robotic systems can make and assemble with pinpoint precision. Faster construction timeframes and higher quality structures are achieved through the use of robotics technology, which automates the assembly process and increases work precision.

Furthermore, robotic technology enables real-time control and monitoring of construction operations, enhancing accuracy and precision on the job site (Asadi et al., 2018). Robots with sensors and other monitoring gear can keep tabs on the building site and flag any discrepancies from the blueprints at any time. With this kind of immediate feedback, project managers may head off problems before they even arise, greatly reducing the possibility of mistakes and setbacks.

Remote Operation and Monitoring Capabilities

Robotics' introduction to the construction industry has revolutionized site management by allowing for remote operation and monitoring (Tijani and Feng, 2021). In contrast to monitoring, which is observing and assessing events as they unfold on a construction site, remote operation entails commanding machinery and tools from afar. These capabilities offer a lot of benefits, including improved safety, increased efficiency, and greater agility in building operations.

With the ability to operate remotely, construction businesses can safely complete operations in areas that are dangerous or difficult to access, protecting human workers from harm (Rasheed et al., 2024). This is absolutely invaluable in fields like emergency response, demolition, and excavation, where worker safety is important. While performing surveys, inspections, or other jobs in potentially hazardous environments, construction companies can substantially profit from utilizing UGVs and UAVs. UAVs equipped with cameras and sensors. For example, they may assess building exteriors, verify the condition of infrastructure, and remotely monitor the progress of construction sites, providing crucial data that would be inaccessible without them (Halder and Afsari, 2023). Similarly, UGVs can safely cross difficult terrains including rubble and debris to carry out search and rescue operations or cleanup sites, hence minimizing the danger to human rescuers.

Robots provide not only remote-control capabilities but also advanced monitoring capabilities, offering significant benefits to construction organizations by providing vital insights into the status and progress of their projects. The use of state-of-the-art sensors, cameras, and data analytics tools enables the continuous tracking of construction operations, surrounding circumstances, and worker well-being.

Collaborative Robotics: Human-Machine Interaction

Collaborative robots are designed to work alongside humans in a shared workspace, collaborative robotics aim to accomplish shared objectives.

Collaborative robots, sometimes known as "cobots" are designed with safety measures that enable them to engage with people without endangering them, in contrast to standard industrial robots that are kept in cages for security reasons. This opens the door to a more collaborative approach to construction, where human workers use their dexterity, instincts, and problem-solving abilities for more complicated tasks, and robots take care of dangerous or repetitive duties.

Advantages of Collaborative Robotics in Construction

Cobots are transforming the construction industry. They improve building projects in several ways including efficiency, safety, and production (Burden et al., 2022). Due to their unique design Cobots can operate alongside humans, fostering communication and collaboration between the two groups. When applied to construction robotics, this cooperative model offers several significant benefits for the industry as a whole.

One of the main benefits of Cobots in construction is the significant improvement in safety. By assisting humans with hazardous tasks, cobots enhance overall worker safety. They can help with welding, assembling, and carrying heavy materials (Pauliková et al., 2021). Cobots can also be programmed to identify and interact with humans. When an employee enters a designated safety zone, the cobots will immediately cease operating, thereby eliminating the risk of accidents and injuries caused by collisions.

Additionally, by enhancing human capabilities and streamlining workflows, collaborative robots increase efficiency and productivity on construction sites. Cobots can precisely and consistently carry out labor-intensive or repetitive tasks, allowing humans to concentrate on more valuable higher-order activities. For instance, Cobots can shorten project durations and decrease manual labor needs in material handling, site cleanup, and prefabrication (Burden et al., 2022). The use of Cobots in modular construction processes can further optimize productivity and reduce costs by assisting with the assembly of prefabricated components as shown in Figure 7 (Reinhardt et al., 2020). Enhanced quality control is yet another benefit of Cobots in the construction industry. Cobots with vision systems and sensors can measure, inspect and check for quality with pinpoint accuracy during construction (Burden et al., 2022). Cobots can help humans with welding, painting and finishing to guarantee that building materials are up to par. The end effect is structures of better quality, less rework, and happier customers.

Safety Measures and Protocols for Human-Robot Collaboration

The safety of human workers engaging with robots is of the utmost importance as Cobots are becoming more and more integrated into building projects. Ensuring a safe and harmonious working environment on construction sites requires safety measures and regulations for human-robot collaboration (Zhang et al., 2023). Included in these steps are a number of components, such as risk analysis, design considerations, training programs, and continuous monitoring.

To assess the risks associated with human-robot collaboration, it is crucial to conduct thorough risk assessments. These assessments evaluate factors such as the complexity of the task, the capabilities of the robot, the layout of the workspace, and the surrounding environment to determine the safety implications for human workers (Awad et al., 2017). Construction firms can mitigate these risks by implementing safeguards to minimize their impact. Cobots enhance safety by reducing the risk of accidents and injuries when working alongside human personnel. Cobots also simplify procedures, which boosts productivity and efficiency. In addition, by carrying out exact inspections and measurements, Cobots help enhance quality control. Cobots as a whole, is a huge step forward in construction innovation, with a plethora of benefits that could completely alter the building and infrastructure industries in the future.

Training programs are necessary to instruct both human workers and robot operators on the proper practices and procedures for working in collaboration with Cobots in a safe manner. The training programs encompass subjects such as robot operation, emergency response methods, and danger detection to guarantee that workers possess a comprehensive understanding of how to safely engage with Cobots (Saad et al., 2024). Additionally, ongoing training and refresher courses are conducted to reinforce safety awareness and keep workers updated on the latest advancements in robot technology.

Real-time surveillance and oversight are essential for guaranteeing the safety of human-robot cooperation on building sites. Construction firms employ monitoring systems, including cameras, sensors, and remote-control interfaces, to supervise and interfere in human-robot interactions, if needed. These monitoring systems allow managers to promptly identify possible safety dangers or malfunctions and promptly implement corrective measures to prevent accidents or injuries. To guarantee the security and welfare of construction workers on job sites, safety procedures and standards for human-robot cooperation are crucial. Construction businesses can build a safe and congenial working environment for human-robot collaboration by conducting rigorous risk assessments, developing thoughtful designs, offering complete training programs, and deploying real-time monitoring and supervisory systems.

Training and Skill Development for Workers in Robotic Environments

As the number of cobots increases in the construction sector, it is essential to implement comprehensive training and skill development programs for workers in robotic environments. The objective of these programs is to provide construction workers with the essential information, skills, and abilities to effectively cooperate with robots (Nguyen and Hadikusumo, 2018). Regular training of employees in maintenance is very important to ensure the effective maintenance and optimal performance of robotic systems in building sites. Workers receive training to perform routine maintenance tasks such as cleaning, lubricating, and inspecting robots. This training aims to minimize mechanical failures and prolong the robots' lifespan. Furthermore, maintenance training encompasses the implementation of troubleshooting procedures with the objective of recognizing and resolving typical challenges or malfunctions that may arise throughout the course of operation (Ali et al., 2024).

Programming training empowers individuals to instruct robots to execute precise tasks and sequences independently. Workers acquire proficiency in programming languages, such as Robot Operating System (ROS) and Python, in order to develop personalized robot behaviors and processes that are specifically designed for building projects. Advanced programming training encompasses subjects such as motion planning, path optimization, and sensor integration. Workers undergo training to recognize potential safety concerns related to robot operation, such as pinch spots, collision zones, and electrical hazards. They then apply suitable safety measures to reduce these risks. The training session covers the procedures for emergency stops, the functioning of safety interlock systems, and the required personal protective equipment (PPE) to ensure compliance with safety standards and regulations (Choi et al., 2022). In addition, training packages tailored for employees in robotic environments emphasize the importance of effective collaboration and communication between humans and robots.

Ethical and Societal Implications of Human-Robot Interaction in Construction

Cobots are profoundly transforming the construction sector, resulting in numerous ethical and social implications that necessitate careful consideration and proactive management. Human-robot interaction (HRI) in construction gives rise to concerns over the replacement of jobs, the independence of workers, safety, privacy, and the wider socio-economic consequences of automation in the constructed environment. The potential for job loss and changes to the labor market is a major ethical concern with HRI in the construction industry (Zaland et al., 2024). Some segments of the construction industry may face employment losses or displacement as a result of robots' growing ability. Fair distribution of jobs, adequate training and skill development for displaced people are all important ethical considerations.

Concerns about human and robot safety when working together in the construction sector are of the utmost ethical importance. Despite the design of Cobots to work closely with humans without posing any harm, accidents and injuries can nevertheless occur. This is especially accurate in the volatile and constantly evolving construction sector. It is important to implement strict safety standards, conduct thorough risk assessments, and continuously supervise and monitor activities 24/7 to minimize possible hazards and avoid accidents involving both humans and robots (Zhang et al., 2023). Data privacy, monitoring, and authorization are some of the concerns brought up by the vast quantities of data gathered by sensor-equipped robots on construction sites. To ensure that all the applicable laws and regulations are followed and to safeguard the privacy of their stakeholders and employees, construction businesses should establish transparent rules and processes for gathering, storing, and utilizing data. A thorough examination and proactive management are necessary in light of the countless ethical and social consequences that arise from HRI in the construction industry. The integration of robotics in construction must promote the common good, respect human dignity and foster a sustainable built environment for all.

Future Directions

Emerging technologies are poised to significantly alter the built environment and redefine human-machine interaction, promising a bright future for cobots in construction. Progress in AI, sensor technology, HRI and modular robotic systems are a few important developments that are influencing the future of cobots. Cobots are led by AI, which empowers robots to do complicated tasks with enhanced intelligence and autonomy. Through self-reflection, course correction, and split-second decision-making, robots programmed with machine learning algorithms can thrive in the unpredictable and ever-changing construction business. By analyzing data from various sensors and cameras, AI-enabled robots can improve their perception of their surroundings, comprehension of human commands, and coordination with people.

Sensor technologies are crucial for enhancing the capabilities of collaborative robots by providing real-time feedback and environmental awareness. Advanced sensors such as LiDAR, 3D cameras, and force/torque sensors enable robots to accurately perceive their surroundings, navigate challenging terrains safely, and interact effectively with both people and structures. By facilitating the quick deployment and reconfiguration of robotic solutions to suit the changing demands of building projects, modular robotic systems are quickly becoming a game-changing technology in collaborative robotics. From material handling and assembly to inspection and maintenance, modular robots can do it all with the help of their interchangeable parts and modules. Modular robotic systems allow construction organizations to scale more easily, adapt to changing project requirements, and improve resource efficiency.

References

Aguair, M. L. and Behdinan, K. (2015). Design, prototyping, and programming of a bricklaying robot. *Journal of Student Science and Technology*, 8(3): 27–34. https://doi.org/10.13034/jsst.v8i3.92.

Ali, Z., Saad, S., Rasheed, K. and Ammad, S. (2024). AI future perspectives and trends in construction. In *AI in Material Science* (pp. 239–261). CRC Press. https://doi.org/10.1201/9781003438489-11.

Ammad, S., Saad, S., Bashir, M. T., Qureshi, A. H., Altaf, M. et al. (2021). Construction accidents via integrating Building Information Modelling (BIM) with emerging digital technologies a review. *2021 3rd International Sustainability and Resilience Conference: Climate Change*. https://doi.org/10.1109/IEEECONF53624.2021.9668066.

Asadi, K., Ramshankar, H., Pullagurla, H., Bhandare, A., Shanbhag, S. et al. (2018). Building an integrated mobile robotic system for real-time applications in construction. *ISARC 2018—35th International Symposium on Automation and Robotics in Construction and International AEC/FM Hackathon: The Future of Building Things*, 96: 470–482. https://doi.org/10.22260/isarc2018/0063.

Awad, R., Fechter, M. and Van Heerden, J. (2017). Integrated risk assessment and safety consideration during design of HRC workplaces. *IEEE International Conference on Emerging Technologies and Factory Automation, ETFA*, 1–10. https://doi.org/10.1109/ETFA.2017.8247648.

Badshah, I., Saad, S., Ammad, S., Rasheed, K., Khan, M. W. et al. (2024). Enhancing sustainable mobility: Lane change strategies for efficient traffic flow at stationary bottlenecks. *2024 ASU International Conference in Emerging Technologies for Sustainability and Intelligent Systems (ICETSIS)*, 1197–1202. https://doi.org/10.1109/ICETSIS61505.2024.10459583.

Bloss, R. (2016). Collaborative robots are rapidly providing major improvements in productivity, safety, programing ease, portability and cost while addressing many new applications. *Industrial Robot: An International Journal*, 43: 463–468. https://doi.org/10.1108/IR-05-2016-0148.

Boya, A., Akinradewo, O., Aigbavboa, C., Ebekozien, A., Ramabodu, M. et al. (2023). Bottlenecks to the implementation of automation and robotics in the construction industry. In *Lecture Notes in Civil Engineering: Vol. 358 LNCE* (pp. 139–149). September. https://doi.org/10.1007/978-3-031-32515-1_11.

Burden, A. G., Caldwell, G. A. and Guertler, M. R. (2022). Towards human–robot collaboration in construction: Current cobot trends and forecasts. *Construction Robotics*, 6(3-4): 209–220. https://doi.org/10.1007/s41693-022-00085-0.

Choi, H. -W., Kim, H. -J., Kim, S. -K. and Na, W. S. (2023). An overview of drone applications in the construction industry. *Drones*, 7(8): 515. https://doi.org/10.3390/drones7080515.

Choi, S. H., Park, K. B., Roh, D. H., Lee, J. Y., Mohammed, M. et al. (2022). An integrated mixed reality system for safety-aware human-robot collaboration using deep learning and digital twin generation. *Robotics and Computer-Integrated Manufacturing*, 73: 102258. https://doi.org/10.1016/j.rcim.2021.102258.

Efimov, A. Y., Gorkavyy, M. A., Gorkavyy, A. I. and Solovev, D. B. (2019). Improving the efficiency of automated precision robotics-enabled positioning and welding. *2019 International Science and Technology Conference "EastConf", EastConf 2019*, 1–5. https://doi.org/10.1109/Eastonf.2019.8725362.

Frank, M., Ruvald, R., Johansson, C., Larsson, T., Larsson, A. et al. (2019). Towards autonomous construction equipment-supporting on-site collaboration between automatons and humans. *International Journal of Product Development*, 23(4): 292–308. https://doi.org/10.1504/IJPD.2019.105496.

Halder, S. and Afsari, K. (2023). Robots in inspection and monitoring of buildings and infrastructure: A systematic review. In *Applied Sciences*, 13(4). https://doi.org/10.3390/app13042304.

Haris, M., Saad, S., Ammad, S. and Rasheed, K. (2024). AI in fabrication and construction. In *AI in Material Science* (pp. 169–192). CRC Press. https://doi.org/10.1201/9781003438489-8.

Haris, M., Saad, S., Ammad, S., Rasheed, K., Aman, H. et al. (2024). Integrating Sustainable practices for effective solid waste management in peshawar: Overcoming challenges and promoting the environment. *2024 ASU International Conference in Emerging Technologies for Sustainability and Intelligent Systems (ICETSIS)*, 1–7. https://doi.org/10.1109/ICETSIS61505.2024.10459445.

Hossain, M. A., Zhumabekova, A., Paul, S. C. and Kim, J. R. (2020). A review of 3D printing in construction and its impact on the labor market. *Sustainability*, 12(20): 8492. https://doi.org/10.3390/su12208492.

Kerber, E., Heimig, T., Stumm, S., Oster, L., Brell-Cokcan, S. et al. (2018, July). Towards robotic fabrication in joining of steel. *ISARC 2018—35th International Symposium on Automation and Robotics in Construction and International AEC/FM Hackathon: The Future of Building Things*. https://doi.org/10.22260/isarc2018/0062.

Khan, M. I., Khan, N., Saad, S. and Khan, M. W. (2024). Artificial Intelligence in construction materials. In *AI in Material Science* (pp. 47–64). CRC Press. https://doi.org/10.1201/9781003438489-3.

Khosravani, M. R. and Haghighi, A. (2022). Large-scale automated additive construction: Overview, robotic solutions, sustainability, and future prospect. *Sustainability*, 14(15): 9782. https://doi.org/10.3390/su14159782.

Kurfess, T. R. (ed.). (2018). *Robotics and Automation Handbook*. CRC Press. https://doi.org/10.1201/9781315220352.

Lattanzi, D. and Miller, G. (2017). Review of robotic infrastructure inspection systems. *Journal of Infrastructure Systems*, 23(3): 4017004. https://doi.org/10.1061/(ASCE)IS.1943-555X.0000353.

Leigh, N. G. and Kraft, B. R. (2018). Emerging robotic regions in the United States: Insights for regional economic evolution. *Regional Studies*, 52(6): 804–815. https://doi.org/10.1080/00343404.2016.1269158.

Liu, Y., Habibnezhad, M. and Jebelli, H. (2021). Brainwave-driven human-robot collaboration in construction. *Automation in Construction*, 124: 103556. https://doi.org/10.1016/j.autcon.2021.103556.

Naeem, U. H., Bilal, M., Syed, F., Rasheed, K., Saad, S. et al. (2022). A multilayer encryption model to protect healthcare data in cloud environment. *2022 International Conference on Data Analytics for Business and Industry, ICDABI 2022*. https://doi.org/10.1109/ICDABI56818.2022.10041708.

Nguyen, H. T. and Hadikusumo, B. H. W. (2018). Human resource related factors and engineering, procurement, and construction (EPC) project success. *Journal of Financial Management of Property and Construction*, 23(1): 24–39. https://doi.org/10.1108/JFMPC-05-2016-0023.

Pan, Y. and Zhang, L. (2021). Roles of Artificial Intelligence in construction engineering and management: A critical review and future trends. *Automation in Construction*, 122: 103517. https://doi.org/https://doi.org/10.1016/j.autcon.2020.103517.

Pauliková, A., Babel'ová, Z. G. and Ubárová, M. (2021). Analysis of the impact of human–cobot collaborative manufacturing implementation on occupational health and safety and the quality requirements. *International Journal of Environmental Research and Public Health*, 18(4): 1–15. https://doi.org/10.3390/ijerph18041927.

Piyush, P., Mohamed, E. and Gabriella, S. K. (2021). Identifying the challenges to adopting robotics in the US construction industry. *Journal of Construction Engineering and Management*, 147(5): 5021003. https://doi.org/10.1061/(ASCE)CO.1943-7862.0002007.

Rasheed, K., Saad, S., Shahzad, L., Ammad, S., Ali, A. and Badshah, I. (2021). Application of gesture data recognition in a human-interactive leap motion sensor chair. 2021 *International Conference on Data Analytics for Business and Industry*, ICDABI 2021. https://doi.org/10.1109/ICDABI53623.2021.9655819.

Rasheed, K., Saad, S., Ammad, S. and Badshah, I. (2023). IoT enhanced solar device management: Integrating remote surveillance and energy efficiency for sustainable operation. *2023 TRON Symposium (TRONSHOW)*, 1–6.

Rasheed, K., Saad, S., Ammad, S. and Bashir, M. T. (2024). Industry 4.0 and construction. In *AI in Material Science* (pp. 65–87). CRC Press. https://doi.org/10.1201/9781003438489-4.

Rasheed, K., Shahzad, L., Saad, S., Khan, H. A., Ahmed, W. et al. (2021). Parking guidance system using wireless sensor networks. *2021 International Conference on Decision Aid Sciences and Application, DASA 2021*. https://doi.org/10.1109/DASA53625.2021.9682213.

Rasheed, K., Waqas, M., Saad, S., Zaland, A., Ali, Z. et al. (2024). Ethical and societal implications of AI in construction. In *AI in Material Science* (pp. 216–238). CRC Press. https://doi.org/10.1201/9781003438489-10.

Rasheed, K., Zaland, A., Saad, S., Ammad, S., Rostami, A. et al. (2024). History of AI. In *AI in Material Science* (pp. 15–46). CRC Press. https://doi.org/10.1201/9781003438489-2.

Reinhardt, D., Haeusler, M. H., London, K., Loke, L., Feng, Y. et al. (2020). CoBuilt 4.0: Investigating the potential of collaborative robotics for subject matter experts. *International Journal of Architectural Computing*, 18(4): 353–370. https://doi.org/10.1177/1478077120948742.

Saad, S., Alaloul, W. S., Ammad, S., Qureshi, A. H., Altaf, M. et al. (2020). Design phase carbon emission prediction using a visual programming technique. *2020 2nd International Sustainability and Resilience Conference: Technology and Innovation in Building Designs*. https://doi.org/10.1109/IEEECONF51154.2020.9319978.

Saad, S., Alaloul, W. S., Rasheed, K. and Ammad, S. (2022). Modeling and simulation of construction cyber-physical systems. In *Cyber-Physical Systems in the Construction Sector* (pp. 88–110). CRC Press. https://doi.org/10.1201/9781003190134-6.

Saad, S., Ammad, S., Rasheed, K., Balbehaith, M., Qureshi, A. H. et al. (2023). Enhancing construction management in restoration through BIM-assisted photogrammetric 3D model accuracy assessment. *2023 International Conference on Sustaining Heritage: Innovative and Digital Approaches (ICSH)*, 100–103. https://doi.org/10.1109/ICSH57060.2023.10482845.

Saad, S., Haris, M., Ammad, S. and Rasheed, K. (2024). AI-assisted building design. In *AI in Material Science* (pp. 143–168). CRC Press. https://doi.org/10.1201/9781003438489-7.

Saad, S., Rasheed, K., Ammad, S., Ali, Z., Zaland, A. et al. (2024). The conclusive role of AI in Industry 4.0 construction materials. In *AI in Material Science* (pp. 262–270). CRC Press. https://doi.org/10.1201/9781003438489-12.

Saad, S., Rasheed, K., Ammad, S., Waqas Khan, M., Zaland, A. et al. (2024). Is AI the architect of tomorrow's materials in the age of Industry 4.0? In *AI in Material Science* (pp. 1–14). CRC Press. https://doi.org/10.1201/9781003438489-1.

Saad, S., Zaland, A., Ammad, S. and Rasheed, K. (2024). Robotics in construction. In *AI in Material Science* (pp. 115–142). CRC Press. https://doi.org/10.1201/9781003438489-6.

Sadhu, A., Peplinski, J. E., Mohammadkhorasani, A. and Moreu, F. (2023). A review of data management and visualization techniques for structural health monitoring using BIM and virtual or augmented reality. *Journal of Structural Engineering*, 149(1): 3122006. https://doi.org/10.1061/(ASCE)ST.1943-541X.0003498.

Schöberl, M., Huber, A., Kreppold, S., Dirnaichner, J., Kessler, S. et al. (2023). Cobot uptake in construction: Embedding collaborative robots in digital construction processes. *Construction Robotics*, 7(1): 89–103. https://doi.org/10.1007/s41693-023-00098-3.

Tijani, B. and Feng, Y. (2021). Social impacts of adopting robotics in the construction industry: A systematic literature review. *Proceedings of the 23rd International Symposium on Advancement of Construction Management and Real Estate*, 668–680. https://doi.org/10.1007/978-981-15-3977-0_51.

Toivonen, T. (2012). *Japan's Emerging Youth Policy*. Routledge. https://doi.org/10.4324/9780203093832.

Wang, B., Hu, S. J., Sun, L. and Freiheit, T. (2020). Intelligent welding system technologies: State-of-the-art review and perspectives. *Journal of Manufacturing Systems*, 56: 373–391. https://doi.org/10.1016/j.jmsy.2020.06.020.

Waqas Khan, M., Saad, S., Ammad, S., Rasheed, K., Jamal, Q. et al. (2024). Smart infrastructure and AI. In *AI in Material Science* (pp. 193–215). CRC Press. https://doi.org/10.1201/9781003438489-9.

Xiao, B., Chen, C. and Yin, X. (2022). Recent advancements of robotics in construction. *Automation in Construction*, 144: 104591. https://doi.org/https://doi.org/10.1016/j.autcon.2022.104591.

Zaland, A., Saad, S., Ahmad, W., Ammad, S., Khan, M. W. et al. (2024). Enhancing mechanical properties of concrete by adding polypropylene fibers. *2024 ASU International Conference in Emerging Technologies for Sustainability and Intelligent Systems (ICETSIS)*, 771–778. https://doi.org/10.1109/ICETSIS61505.2024.10459663.

Zaland, A., Saad, S., Rasheed, K. and Ammad, S. (2024). Automation in construction industry. In *AI in Material Science* (pp. 88–114). CRC Press. https://doi.org/10.1201/9781003438489-5.

Zhang, M., Xu, R., Wu, H., Pan, J., Luo, X. et al. (2023). Human–robot collaboration for on-site construction. *Automation in Construction*, 150: 104812. https://doi.org/10.1016/j.autcon.2023.104812.

Chapter 3

Building the Digital Twin

Data Collection and Integration

Muhammad Waqas Khan,[1] *Syed Saad*,[2]
Kumeel Rasheed[1] and *Syed Ammad*[3,*]

Introduction

In recent years, the concept of the "digital twin" has significantly impacted several industries, transforming how we organize, oversee, and enhance technical infrastructure. Digital twins have become essential tools for understanding, simulation, and managing real-world assets and processes across sectors such as manufacturing, smart cities, healthcare, and transportation. As we stand on the brink of the Fourth Industrial Revolution, organizations aiming to thrive in an increasingly digitalized world must integrate digital twins into their daily operations (Rasheed et al., 2024).

This chapter delves into the fundamental shift, emphasizing the crucial aspects of data integration and collection that facilitate the creation and utilization of digital twins. By leveraging the latest research and industry practices it aims to guide practitioners, researchers, and decision-makers through the complexities of deploying digital twins. The swift adoption of digital twins is driven in part by the massive amount of data generated by sensors, Internet of Things (IoT) devices, and other sources.

[1] Department of Civil Engineering, CECOS University of IT and Emerging Sciences, Pakistan.
[2] Science, Engineering & Manufacturing Centre Sandwell College West Bromwich, United Kingdom.
[3] School of Engineering and Built Environment, Griffith University, Gold Coast, Australia.
* Corresponding author: syed.ammad@griffithuni.edu.au

This abundance of data has advantages as well as disadvantages. Predictive maintenance, real-time optimization, and enhanced decision-making are made feasible by the previously unattainable insights into the internal workings of systems that it offers (Achouch et al., 2022). The enormous volume, speed, and diversity of data make processing, storing, and interpreting it extremely complex. Confidentiality, security, and data integrity must still be upheld, particularly in vital sectors like healthcare and banking. The ability to fully utilize digital twins has been made possible by advancements in data integration methods that have kept up with the volume of data generated.

Compatible platforms and standards unify previously isolated data sources, facilitating smooth communication and cooperation across organizational boundaries (Otto and Jarke, 2019). Digital twins can adapt and change in real-time because machine learning and artificial intelligence (AI) algorithms are important in obtaining meaningful insights from diverse data streams (Haris et al., 2024). In addition, the arrival of cloud computing and distributed architectures has enabled digital twins to function at a larger scale, in closer range to the data-generating source, therefore decreasing latency and improving responsiveness.

A diverse effort like creating a digital twin requires careful coordination between domain specialists, data scientists, engineers, and stakeholders. It is more than just a technical task. For this reason, the chapter bridges the gap between theories and practice as well as between academia and industry through an interdisciplinary approach to give readers the knowledge and tools they need to begin their digital twin journeys by offering case studies, best practices, and helpful guidance.

The chapter covers a wide range of topics, including:

- Definitions, guiding concepts, and architectural considerations for digital twins.
- Techniques for gathering data: wearables, sensor networks, IoTs devices, and more.
- Data lakes, APIs, and interoperability standards are examples of data integration strategies.
- Machine learning and data analytics: optimization methods, anomaly detection, and predictive modeling.
- Threat vectors, encryption methods, and regulatory compliance are all aspects of security and privacy.
- Case studies and their uses from precision farming and personalized medicine to intelligent manufacturing and predictive maintenance.

Data Sources for Digital Twins in Construction

The integration of digital twins into the construction industry has fundamentally transformed our approach to project management. This groundbreaking technology relies on the meticulous collection and integration of data from various sources. In this section, we examine the diverse array of data sources that are necessary to build reliable digital twins in the construction sector.

Data Streams from IoT Sensors and Geospatial and GIS Data

The incorporation of data streams from IoT sensors implanted in physical assets is a fundamental aspect of digital twin development in the construction industry. These sensors are scattered over construction sites and equipment and gather a multitude of real-time data on structure integrity, temperature, humidity, equipment operation, and even worker activity (Tang et al., 2019). By utilizing this data, stakeholders may track developments, spot inefficiencies, and improve workflows instantly by having access to a complex digital representation of the real world.

Within the digital twin framework, geographic data enhanced by Geographic Information Systems (GIS) is essential for improving the contextual knowledge of construction projects. Light Detection and Ranging (LiDAR) scans, topographical maps, satellite imagery, and aerial surveys all provide priceless insights into the topography, spatial linkages, and environmental variables (Brown et al., 2016). By identifying possible risks and allocating resources as efficiently as possible, geospatial data integration not only makes precise site design and logistics possible, but it also promotes proactive risk management.

BIM and As-Built Models and Construction Management Systems

Building Information Modelling, also known as BIM, provides complete 3D models enhanced with metadata during the construction lifetime, which is the foundation of digital twins. MEP (mechanical, electrical, plumbing) systems, material requirements, architectural designs, and structural elements are all included in these models (Rodriguez et al., 2020). As construction progresses BIM models evolve into as-built representations, accurately reflecting the project's current state. Digital twins that integrate BIM and as-built models ensure a seamless transition from design to construction, facilitating stakeholder collaboration and supporting informed decision-making (Saad et al., 2023).

The creation of digital twins necessitates a wealth of operational data, which is produced by construction management systems, including project management software, scheduling tools, and enterprise resource planning (ERP) platforms (Lim et al., 2020). By capturing project schedules, resource usage, financial restrictions, and regulatory compliance, these systems provide stakeholders with a thorough picture of the effectiveness and productivity of their projects. Stakeholders can predict project outcomes, reduce delays, and allocate resources optimally under project objectives by integrating data from construction management systems.

Emerging Trends in Data Integration

Some new developments in data integration are changing the building industry as digital twin technology develops further. The convergence of AI and machine learning algorithms has enabled digital twins to predict future events, allocate resources optimally, and proactively mitigate risks (Fan et al., 2021). Moreover, advancements in cloud and edge computing have equipped digital twins with scalable storage, real-time processing capabilities, and improved accessibility, eliminating the limitations associated with data administration and analysis.

In the construction sector, creating reliable digital twins begins with the thorough collection and integration of various data sources. Every data stream, from BIM models and construction management systems to IoT sensors, adds to the comprehensive digital depiction of construction projects (Valinejadshoubi et al., 2022). Stakeholders can fully utilize digital twins to maximize efficiency, improve decision-making, and spur innovation across the construction lifecycle by adopting new trends in data integration.

Sensors and IoT Devices: The Neural Network of Digital Twins

The IoT devices and a network of sensors enable the smooth integration of data gathering and utilization, which is at the core of this digital revolution. This section explores the complex network of sensors and IoT devices, explaining their essential function as the brain network of digital twins in the construction industry. Understanding digital twins in the context of construction is essential before delving into the nuances of sensors and IoT devices. A virtual copy of real assets, procedures, or systems is known as a digital twin. In the context of construction, this means constructing a digital replica of infrastructure, structures, or even whole construction sites. Throughout the project lifecycle, real-time monitoring, analysis, and optimization are made possible by this digital duplicate.

Sensors' Vital Role in Data Gathering and Spatial Awareness Enhancement

Sensors are the main source of data capture in the complex field of building design and construction, acting as essential entry points between the digital and physical domains (Saad et al., 2024). These modest but potent instruments have an amazing capacity to record a wide range of characteristics, from minute vibrations and stresses in the structure to delicate variations in humidity and temperature. Sensors are essential at every stage of a building project's life cycle, from the first phase of material acquisition and site assessment to continuous structural monitoring and maintenance (Zaland et al., 2024). A key component of this sensor technology is the generation of accurate 3D models that represent assets and construction sites. Advanced spatial sensors like photogrammetry systems and LiDAR scanners provide this level of precision. Together, LiDAR scanners and photogrammetry systems generate comprehensive spatial maps by using quick laser pulses and photographic images, respectively (Zhang and Lin, 2017). These maps provide stakeholders with an unmatched degree of accuracy in terms of interpreting and visualizing complex structures. From the minutest characteristics of a building's surroundings to the subtlest nuances in its contours, these sensors provide architects, engineers, and project managers with a comprehensive view, as illustrated in Figure 1.

These spatial sensors play a crucial role in enabling the creation and functioning of digital twins, which are virtual replicas of real-world buildings. Digital twins can track building progress in real time due to the continuous stream of data provided by these sensors. They have the ability to quickly identify changes from the original design, allowing for the preservation of project budgets and schedules. Additionally, data

Figure 1: Exploring the sensor network in Digital Twins.

from spatial sensors aids in optimizing spatial layouts for enhanced effectiveness, ensuring that every square inch of a structure is utilized to its fullest potential. In simple terms, the introduction of sensors into the building ecosystem signifies a radical change in the way we design, build, and maintain buildings (Saad et al., 2024). They not only bridge the gap between digital planning and physical construction, but also herald a new era of sustainability, efficiency, and precision in the built environment. As technology continues to advance, the role of sensors will undoubtedly expand, revolutionizing building and design with each new innovation.

Sensing Sustainability & Safety: Structural Health Monitoring

As construction projects revolve around maintaining structural integrity, ongoing observation and evaluation are crucial. Buildings, bridges, and other infrastructure can have their structural integrity assessed with the use of sensor networks that include accelerometers, displacement sensors, and strain gauges. These sensors enable preventive maintenance interventions and reduce the danger of major breakdowns by detecting minute variations in load distribution, material qualities, and ambient conditions (Abdulkarem et al., 2020; Moreno-Gomez et al., 2018). The use of environmental sensors becomes critical in a time when safety regulations and environmental awareness are becoming more and more important. Hazardous gas detection, noise level measurement, and air quality monitoring are just a few of the many functions that these sensors can perform (Shahzad et al., 2021). Environmental sensors enable digital twins to promote safer construction methods, reduce environmental impact, and guarantee regulatory compliance by tracking ambient conditions and identifying possible dangers.

Visualizing Environmental Sensor Functionalities

In the modern world, when environmental awareness and safety laws are becoming more and more important, using environmental sensors becomes essential. These sensors perform a wide range of tasks, such as monitoring air quality, assessing noise levels, and identifying dangerous substances. Their function in this domain goes well beyond simple data gathering; they serve as the foundation for the creation of digital twins. Environmental sensors assure compliance with regulations, reduce environmental impact, and open the door for safer construction practices. These sensors act as vigilant defenders of environmental stewardship and safety, swiftly and accurately detecting potential threats. Their integration into diverse industries signifies a noteworthy advancement towards a future that is both secure and sustainable (Haris et al., 2024), where technology advancements are in harmony with the health of our planet. This paradigm shift towards sensor-driven environmental awareness offers a deep dedication to ethical and responsible behavior in addition to improved safety procedures. Figure 1 underscores the crucial role these sensors will play in ensuring industries operate in compliance with safety regulations and environmental protection standards.

IoT Devices: Data Fusion, Intelligence, and Sensor Integration

The advent of IoT devices heralds a new era in data communication and intelligence augmentation. These networked devices take various forms, including smart cameras, actuators, wearables, and drones. Digital twins rely on IoT devices for adaptive control mechanisms, real-time analytics, and seamless data transfer, essential for their functionality (Khan et al., 2024). By incorporating IoT devices into their static representations, digital twins become dynamic cyber-physical systems that are capable of independent decision-making and environment response. Although sensors and IoT devices hold great promise to enhance digital twins, numerous barriers must be overcome to achieve seamless (Rasheed et al., 2023). These challenges include scalability, cybersecurity, interoperability, and data standards issues. All obstacles, however, present opportunities for ingenuity and collaboration.

In the context of contemporary technology, sensors and IoT devices are the key components that propel the development of digital twins. These digital twins, which mimic real-world resources or operations, are cognitively supported by sensors and IoT devices, which facilitates the smooth interaction of data gathering, processing, and useful application. Stakeholders in a variety of industries are freeing themselves from the limitations of conventional construction methods by implementing spatial sensors, environmental monitors, IoT devices, and structural monitors (Rasheed et al., 2024). With this freedom comes a new era characterized by sustainable progress, operational excellence, and data-centric decision-making.

Organizations can overcome traditional constraints by harnessing the capabilities of these numerous sensors, paving the way for an era of innovation built on a foundation of accurate, up-to-date data. Geospatial sensors provide rich information on the spatial organization of areas, facilitating effective resource distribution and maximizing use. Environmental monitors keep a close eye on variables like temperature and air quality to guarantee not just comfort but also health and safety (Song et al., 2019). IoT devices, in all forms serve as the glue that holds together an intelligent network that extends from factories to intelligent homes (Rasheed et al., 2022).

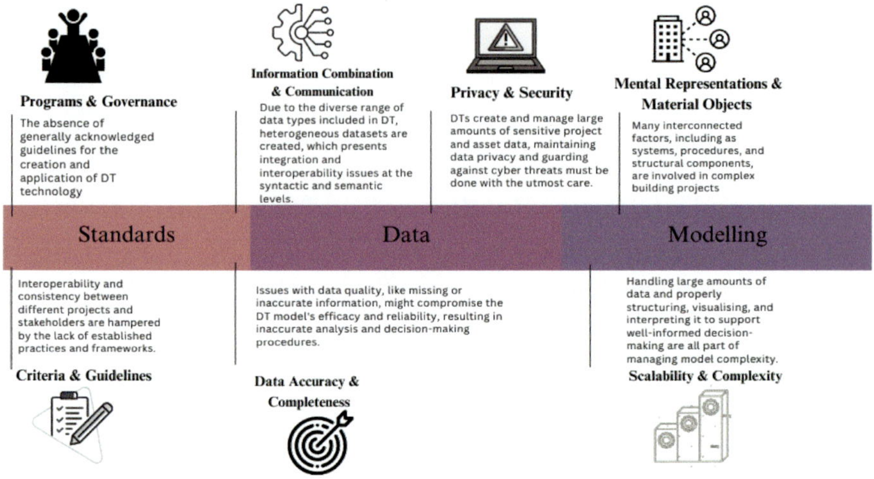

Figure 2: Navigating data models and standards in the IoT universe.

Furthermore, the incorporation of structural monitors into this technology framework enables stakeholders to track asset performance and health in hitherto unheard-of ways. These sensors work as watchful sentinels, delivering early warnings and enabling proactive maintenance, whether they are monitoring the stress levels of a bridge, the structural integrity of a building, or the effectiveness of industrial gear (Saad et al., 2024). This combination of sensing technologies is a reinvention of what is possible, not just about efficiency as shown in Figure 2.

Embarking on this journey of sensor integration leads into a realm that encourages innovation and creativity, rather than simply adopting a new technology. Digital twins have the ability to completely transform a variety of businesses by providing an unprecedentedly precise canvas on which ideas can be tested, improved, and put into practice (Cooper et al., 2022).

The story in this changing environment moves from being reactive to being proactive and from conjecture to data-driven insights. Numerous real-time data points inform every decision, be it optimizing supply chains or designing a building with higher energy efficiency (Rasheed et al., 2023). In this future, efficiency issues are identified and resolved before they become noticeable, and sustainability is achieved through well-considered decisions rather than only a buzzword. So, while we work through the challenges of sensor integration, let's embrace the revolutionary promise of digital twins. Let us take advantage of this chance to design a future in which the buildings we construct, the settings we live in, and the technologies we depend on are not only works of fiction but also representations of a sustainable, data-driven culture.

Data Integration and Management

The concept of a digital twin has revolutionized the construction industry, changing traditional approaches to project management, planning, and execution. The digital twin serves as the virtual counterpart of a physical asset, offering real-time data and enabling informed decision-making throughout the asset's life cycle. However, the successful implementation of a digital twin depends on the smooth administration and integration of heterogeneous datasets from different sectors of the building industry. Data integration and management are fundamental to the digital twin because they enable the aggregation, harmonization, and analysis of diverse data sources (Jeddoub et al., 2023). This section explores the complexities of data management and integration in the context of creating the digital twin for the construction sector. The ability of a digital twin project to compile, integrate, and analyze this vast array of heterogeneous data sources is fundamental to its success. This highlights how crucial data management and integration are, since they essentially serve as the framework for building the digital twin. This process has many inherent complications, from the difficulty of assembling data from several project life cycle stages to making sure that diverse stakeholders' technologies and data formats are compatible.

In the context of digital twins, data integration involves synthesizing data from various sources, including sensors, IoT devices, BIM, GIS, and more. These sources offer a wealth of data covering everything from material composition and structural integrity to user behavior and patterns of energy consumption. But the intrinsic uniqueness and diversity of these datasets present a significant hurdle, necessitating the use of advanced techniques and tools in order to successfully align and standardize the data (Costa et al., 2024).

The dynamic nature of building projects, characterized by constantly changing specifications, designs, and unpredictability, significantly complicates data management (Ammad et al., 2021). In order to guarantee that the digital twin continues to be a precise representation of its physical counterpart, real-time updates and data synchronization become essential. In order to build confidence and compliance among stakeholders, it is also necessary to carefully manage concerns related to data security, privacy, and ownership.

Understanding Data Integration

Data integration is the process of combining information from several sources to create a unified, consistent picture for analysis and decision-making. This refers to integrating data from various sources, including facility managers, suppliers, contractors, engineers, and architects. The goal is to create a comprehensive digital representation of the physical asset and its associated operations. Several issues, such as data heterogeneity, inconsistency, and interoperability, must be resolved to achieve seamless data integration (Ali et al., 2024). Several different kinds of data are produced by construction projects: sensor readings, project timelines, BIM, and geospatial data. Integrating these datasets can be quite difficult because they frequently follow disparate formats, standards, and conventions. Construction stakeholders need to embrace standardized data formats and protocols (Saad et al., 2024), such as GIS standards for geospatial information and Industry Foundation Classes (IFC) for BIM data, in order to overcome these obstacles. Furthermore, the establishment of data exchange protocols and interoperability frameworks enables smooth system integration and communication. Complexity is increased by the wide variety of data produced by construction projects, including sensor readings, BIM, and geographic data (Barzegar et al., 2021). Integration of these datasets can be challenging since they frequently have distinct forms, adhere to different standards, and follow different practices. Construction stakeholders need to adopt standardized data formats and protocols in order to overcome these obstacles. For example, frameworks for compatibility are provided by IFC for BIM data and GIS standards for geographic information. Establishing interoperability frameworks and data exchange protocols is also essential for efficient system integration and communication throughout the construction ecosystem (Saad et al., 2020). The construction sector may improve decision-making, project efficiency, and overall operational performance by utilizing data integration through these initiatives.

Data Management Strategies

Throughout its existence, construction data must remain accessible, secure, and of high quality. This requires effective data management. To do this, strong data governance frameworks, methods for ensuring data quality, and data lifecycle management procedures must be established. The policies, processes, and roles controlling the gathering, storing, and use of construction data are outlined in data governance frameworks. Delineating data ownership, access controls, and data stewardship practices, these frameworks guarantee adherence to industry standards and regulatory mandates. Construction stakeholders can reduce the risks of data breaches, abuse, and unauthorized access by putting in place clear governance frameworks. In order to guarantee the precision, comprehensiveness, and dependability of construction data, data quality assurance procedures are essential. Processes for quality assurance include data validation, cleansing, and enrichment methods to identify and resolve mistakes, discrepancies, and abnormalities. The effectiveness of data quality assurance initiatives is increased by utilizing automated data validation tools, machine learning algorithms, and domain-specific validation criteria (Zaland et al., 2024), which raises the digital twin's overall level of trustworthiness. Data lifecycle management practices control the entire process of managing construction data, from creation to archiving or disposal. This includes developing data retention policies, backup and recovery methods, and metadata management strategies to improve data traceability, auditability, and compliance. Construction stakeholders can improve storage utilization, reduce data redundancy, and assure regulatory compliance by following strict data lifecycle management standards.

Data Management and Investigations

A thorough method of managing data effectively for research or analysis is shown in Figure 3. The process begins with the essential step of gathering information required from reliable and relevant sources, which lays the foundation for the entire procedure. Once the data is collected, the focus shifts to identifying and retaining key data points pertinent to the ongoing study. This systematic approach helps avoid distractions from irrelevant data, ensuring that resources are concentrated on the most pertinent information. Figure 3 underscores the importance of combining data from different sources or industries. Investigations often require data from multiple surveys, different departments within an organization, or various geographic areas. Establishing linkages and combining various datasets is essential to completing the picture and producing meaningful insights.

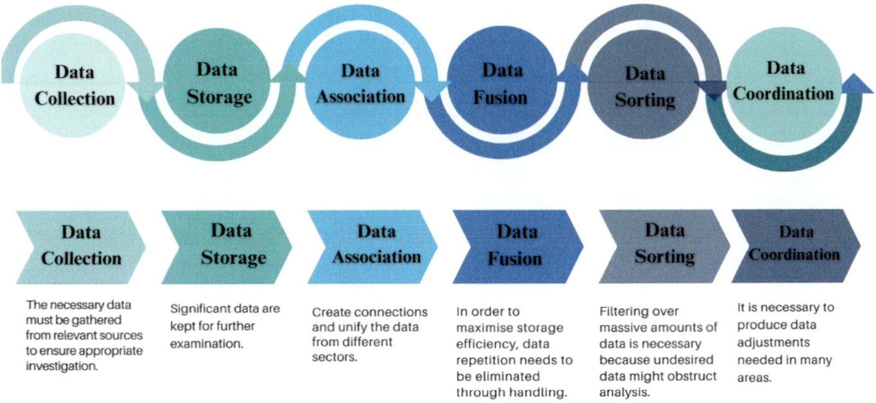

Figure 3: Data analysis process for building a successful digital twin.

When working with big datasets, efficiency is crucial. Eliminating data repetition, which involves identifying and storing duplicate information is one such technique. This reduces the need to sort through duplicate records, which speeds up analysis procedures in addition to saving storage space. When working with enormous volumes of data, filtering becomes essential. Eliminating unnecessary information allows the analysis to concentrate on the most pertinent and helpful data, producing results that are more accurate (Khan et al., 2014).

Figure 3 also highlights the requirement for data modifications. Certain data may require modifications in order to clean, standardize, or alter it because not all of it is flawless or instantly usable. This could entail standardizing data from several sources, fixing mistakes, or transforming data formats. These modifications guarantee the precision and coherence of the facts utilized throughout the inquiry.

Best Practices in Data Integration and Management

Optimizing data integration and management processes for the creation of a digital twin involves several best practices that can be highly beneficial to construction stakeholders. First and foremost, it is crucial to set specific goals. Stakeholders need to specify their goals before starting the process of building a digital twin (Rasheed et al., 2024). This could entail optimizing team cooperation, increasing operational efficiency, or simplifying maintenance procedures. The data integration process gains direction and purpose when these objectives are kept in mind.

Second, it is critical to choose the appropriate data sources. Massive volumes of data are produced throughout construction projects by a variety of systems, including project management software, IoT sensors, BIM, and more. In order to prevent data overload and ensure the digital twin accurately represents the most important features of the physical asset, it is important to determine which data points are vital to the objective of the twin (Javaid et al., 2023). Third, it is crucial to ensure data quality throughout the integration process. The efficacy of the digital twin can be significantly impacted by inconsistent or inaccurate data. Putting in place procedures for data validation and conducting routine audits can support the upkeep of data integrity, offering a solid basis for making decisions based on the digital twin. Another important factor that cannot be disregarded is security. Cyber dangers must be avoided when handling construction data, particularly when dealing with private information like building plans and project schedules (Oesterreich and Teuteberg, 2016). Strong cybersecurity measures, such as access controls and encryption, are put in place to protect the digital twin and the priceless data it contains.

In order to fully realize the promise of the digital twin, project teams must cultivate a culture of data exchange and use. Educating staff members on how to work with and understand data from the twin gives them the ability to make wise decisions. Throughout the project lifecycle, encouraging cross-functional collaboration guarantees that insights from the digital twin are utilized across departments, fostering efficiency and creativity.

Crafting a Data Integration Roadmap and Enacting Agile Integration Methods

A comprehensive plan is developed to define the goals, milestones, and timescales for data integration projects, with collaboration from key stakeholders to prioritize activities based on business goals and strategic objectives. Agile approaches are adopted for iterative data integration and deployment, allowing for rapid feedback, iteration, and adaptation. Data integration efforts are segmented into smaller, more manageable components to facilitate incremental development and continuous improvement. In investing in interoperability technologies and standards, organizations aim to facilitate seamless data flow and integration across multiple systems and platforms. Utilizing open-source data integration tools, APIs, and data exchange standards enhances interoperability and data portability (Borgogno and Colangelo, 2019). A key strategy in data integration projects involves breaking them down into smaller parts to enable incremental development, minimize the risk of major setbacks, and promote ongoing improvement. This approach ensures each stage of integration is thoroughly evaluated and optimized before progressing, thereby enhancing overall reliability and quality of data movement between platforms and systems. This entails using APIs, implementing open-source data integration technologies, and following data exchange guidelines.

Cost-effectiveness and flexibility are provided by open-source tools, while the required interfaces for successful system communication are provided via APIs. Data formats must be compatible and consistent for interoperability to be further improved by data exchange standards (Borgogno and Colangelo, 2019). This all-encompassing strategy improves data portability and integration efficiency, facilitating the effortless transfer and utilization of data across several platforms and applications. By implementing these tactics, businesses can create strong, flexible frameworks for data integration that meet their present requirements and provide a strong basis for expansion and innovation in the future.

Facilitating Data Collaboration, Sharing, and Enforcing Governance Protocols

To foster a culture of data collaboration and sharing among construction stakeholders, emphasis is placed on interdisciplinary collaboration and information exchange. Partnerships, consortiums, or agreements are formed to facilitate data exchange while safeguarding confidentiality and intellectual property rights. Strong data governance procedures are established to ensure data integrity, security, and compliance. This includes defining roles, responsibilities, and data governance principles. Systems are implemented to monitor and enforce adherence to data governance standards, ensuring consistency and reliability in data management practices across the organization (de Streel et al., 2021). Leverage advanced technologies such as AI, ML, and the IoT to improve data integration and management capabilities. Use IoT devices for real-time data collection and monitoring, apply machine learning algorithms for data enrichment and purification, and implement AI-driven data integration solutions. These technologies work together to enhance data accuracy, generate insights, and boost operational efficiency across various applications.

The building blocks of the digital twin are data integration and management, which allow for the collection, combining, and analysis of varied information for well-informed decision-making. By implementing best practices in data integration and management, construction stakeholders may overcome the issues of data heterogeneity, inconsistency, and interoperability, allowing the digital twin to fully revolutionize the construction sector. Collaborative efforts, technological innovation, and a dedication to data governance may realize the vision of a seamlessly networked digital twin for construction projects, starting in a new era of efficiency, sustainability, and resilience in the built environment.

Security and Privacy Concerns in Digital Twin Data

In digital twin technology, which replicates physical entities in a virtual environment to enhance decision-making and operational efficiency, security and privacy are significant concerns. As digital twins become more integrated across various industries addressing these issues becomes increasingly urgent. The primary concern is the susceptibility of digital twin systems to cyberattacks and breaches. Sensitive data is hidden within these exact duplicates of real-world assets, which range from industrial gear to complete smart cities (Jeddoub et al., 2023). Malicious hackers could exploit vulnerabilities in these systems, jeopardizing not only the integrity of the twin but also the physical entity it represents. Such intrusions could have disastrous repercussions, such as compromised security in vital infrastructure and interrupted operations in manufacturing facilities (Waqas Khan et al., 2024). Moreover, the legal framework pertaining to digital twin security and privacy is still in its early stages and is not well-defined. Regulations and laws pertaining to cybersecurity and data protection frequently find it difficult to keep up with the quick advancement of technology. Digital twin adopters face a number of difficulties as a result of this regulatory lag, including uncertainty about their legal responsibilities and liabilities.

Organizations need to take a multifaceted approach to strengthen the security and privacy of their digital twins in order to navigate these dangerous seas. First and foremost, to protect data while it is in transit and at rest, strong encryption techniques should be used. Strict access restrictions that use multi-factor authentication and restrict access to authorized workers are required. It is essential to conduct regular audits and vulnerability assessments in order to find and fix weak points early.

Understanding Security in Digital Twin Data

Threats to digital twin data security are numerous and range from cyberattacks and data breaches to unauthorized access. Information integrity and confidentiality are particularly challenging to protect when dealing with digital twins because of their inherent nature, which consists of constant real-time data transmission between virtual and physical counterparts. Data authorization and authentication are two main areas of concern. It is crucial to make sure that only authorized individuals have access to sensitive data because digital twin systems are interconnected. The integrity of the entire system may be compromised if strong authentication measures are not put in place. Furthermore, since digital twins frequently combine data from several sources, such as cloud platforms and IoT devices, protecting data transfer channels is essential to avoiding malicious actors' manipulation and interception. Safeguarding information is an important component of security with digital twin data. This entails using encryption methods to protect data when it's in motion or stationary. Encryption reduces the possibility of data leakage or exploitation by guaranteeing data confidentiality and preventing unwanted access to sensitive information. Sustaining operational continuity also requires protecting digital twin infrastructure against cyber threats like ransomware, malware, and distributed denial-of-service (DDoS) attacks (Naeem et al., 2022). Organizations can mitigate the risk of cyber-attacks and minimize information compromises and system outages by implementing firewalls, intrusion detection technologies, and periodic security audits as shown in Figure 4.

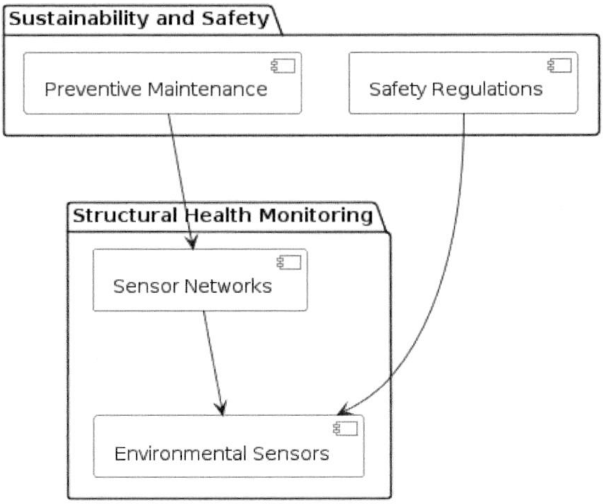

Figure 4: Digital Twins for structural health monitoring and sustainability management.

Addressing Privacy Concerns in Digital Twin Data

Privacy concerns in digital twin data mostly center on the gathering, storing, and use of private or sensitive data. Adhering to data protection standards becomes crucial as digital twins collect copious amounts of data from several sources, such as environmental sensors, operational logs, and user interactions. A major obstacle is pseudonymization and anonymizing data without sacrificing its usefulness. Analyzing personally identifiable information (PII) in an anonymous manner preserves user privacy, but it may also reduce the efficiency of data analysis and decision-making. Deriving useful insights from digital twin data while protecting individual privacy rights requires striking a balance between data anonymization and utility. Moreover, in digital twin ecosystems, consent management becomes a crucial component of privacy regulation. Upholding privacy principles and legal requirements requires getting an explicit agreement from data subjects for the collection, processing, and dissemination of their personal information (Vishwamitra et al., 2022). Within the digital twin ecosystem, transparency in communication about data usage regulations and giving consumers fine-grained control over their data can promote trust and compliance.

Moreover, consent management appears as a crucial privacy control element inside the intricate ecosystems of digital twins. Obtaining explicit consent from data subjects on the collection, processing, and dissemination of their personal information is necessary to uphold privacy values and legal requirements. It is imperative that communication policies pertaining to data usage within the digital twin ecosystem are transparent. Giving customers precise control over their data helps promote compliance and confidence. Users are more likely to feel empowered and confident while interacting with digital twin technologies when they are aware of how their data will be used and have the ability to make informed decisions about sharing it.

This openness guarantees that privacy rules are respected for the benefit of all parties involved and builds confidence between users and data collectors. Prioritizing privacy and putting strong consent mechanisms in place will be essential for navigating the ethical and legal challenges that come with using such cutting-edge technology as the landscape of digital twins continues to change.

Compliance with Data Protection Regulations

Organizations have to manage a complicated regulatory environment that controls the handling and collecting of digital twin data in addition to technical security and privacy issues. Tight guidelines on data protection, accountability, and transparency are placed on data controllers and processors by laws like the California Consumer Privacy Act (CCPA) in the US and the General Data Protection Regulation (GDPR) in the EU (Baik, 2020). It takes strong data governance frameworks, privacy impact analyses, and the appointment of data protection officers to manage compliance initiatives to achieve compliance with these standards. To guarantee legal data processing and transfer procedures, businesses with international operations must also handle the complexities of international data transfer methods like the EU-US Privacy Shield and standard contractual clauses. Strong data governance frameworks must be established and maintained by organizations in order to comply with these strict demands. These guidelines provide the foundation for making sure that the collection, storage, and use of digital twin data adheres to legal standards. Furthermore, privacy affect evaluations are essential for spotting any dangers and weak points in the data processing lifecycle. Organizations can proactively address privacy risks and install required measures by undertaking these analyses. The employment of data protection officers (DPOs) is a crucial part of compliance measures. In addition to guaranteeing internal compliance and acting as a point of contact for data protection authorities, DPOs are responsible for managing data protection initiatives (Piasecki, 2023). Their function is essential to establishing a privacy and data protection culture within the organization. Furthermore, the difficulties in guaranteeing lawful data processing and transfer protocols are exacerbated for companies with global operations. There are added complications when using international data transfer techniques like the EU-US Privacy Shield and standard contractual agreements.

Best Practices for Mitigating Security and Privacy Risks

To address security and privacy concerns related to digital twin data, it is recommended that organizations take a comprehensive approach that integrates organizational, technical, and constitutional protections. Among the finest practices are:

- Adopting a zero trust security paradigm, which does not just rely on perimeter defenses but also enforces access controls based on user identification, device posture, and contextual variables.
- Using strong encryption methods and key management procedures to safeguard the confidentiality and integrity of data at all stages of its existence.
- Regularly carrying out penetration tests and vulnerability assessments to find and fix security flaws in infrastructure and digital twin systems.
- Making sure that privacy concerns are ingrained from the beginning by including privacy by design principles into the creation and implementation of digital twin solutions.
- Offering thorough user education and training courses to inform staff members, partners, and stakeholders on security and privacy best practices.

The broad adoption and application of digital twin technology across numerous industries is severely hindered by security and privacy issues. Corporations can proactively handle security threats, privacy risks, and legal requirements and foster stakeholder trust by being aware of the subtleties of these issues (Zanke et al., 2024). Realizing the full potential of digital twins while safeguarding data integrity and individual privacy rights necessitates adopting a multifaceted strategy. This strategy should encompass technological safeguards, privacy enhancing technologies, and rigorous regulatory compliance procedures.

Case Studies: Creating Comprehensive Digital Twins

In this section of the chapter, the focus is on investigating the real-world applications of digital twins across various sectors. The investigation focuses on how the integration of data collection techniques with the development of digital twins produces revolutionary results across multiple in-depth case studies. Each research emphasizes how data collecting methods may be seamlessly integrated, and it also shows how creating and using digital twins can yield previously unheard-of insights and efficiencies. The exploration of these real-world applications provides insights into how this technology moves beyond theoretical frameworks to become a cornerstone of contemporary industry. Digital twins are changing the approach of problem-solving and decision-making, whether it be by improving predictive maintenance, revolutionizing urban planning, or optimizing production processes. Through these insightful case studies, we are able to personally observe the critical role that digital twins play in fostering innovation and advancement across a range of industries, leading us closer to a day when the digital and physical worlds will smoothly merge for improved results.

Healthcare: Revolutionizing Patient Care

Digital twins are invaluable for both patients and healthcare providers in the healthcare industry, extending far beyond mere replication. Healthcare workers create digital replicas that capture each individual's distinct physiology and health profile by carefully integrating patient data, medical records, and diagnostic imaging. AI is still in its early stages of being fully utilized for medical diagnosis. However, more data is emerging for the application of AI in diagnosing different diseases, such as cancer. A study was published in the UK where authors input a large dataset of mammograms into an AI system for breast cancer diagnosis. This study (McKinney et al., 2020) showed that utilizing an AI system to interpret mammograms had an absolute reduction in false positives and false negatives by 5.7% and 9.4%, respectively. Another study was conducted in South Korea, where authors compared AI diagnosis of breast cancer versus radiology. The AI-utilized diagnosis was more sensitive to diagnose breast cancer with mass compared to radiology, 90% vs. 78%, respectively (Choi et al., 2022). Also, AI was better at detecting early breast cancer (91%) than radiology's 74%.

Furthermore, a study utilized deep learning to detect skin cancer which showed that an AI using Convolutional Neural Networks (CNN) accurately diagnosed melanoma cases compared to dermatologists and recommended treatment options (Naqvi et al., 2023). This feature of digital twins is especially significant since it improves patient care by enabling prompt reactions to changes in health. Digital twins in healthcare, as a whole, signify a profound change in the way that healthcare is provided. They constitute a paradigm shift towards proactive and individualized healthcare, not only diagnostic or therapeutic instruments. The real world applications of digital twins underscore their potential to revolutionize the healthcare industry, offering a glimpse into a future where patient centered, accurate, and efficient treatment becomes standard practice.

Manufacturing: Enhancing Operational Excellence

In the realm of aerospace innovation, digital twins' examples have emerged as stalwart companions, intricately mirroring the complex aircraft systems. These digital counterparts open up a wealth of possibilities, enabling meticulous analysis and prognostication within the domain of aircraft performance. At the core of this aerospace revolution lies the continuous vigilance over engine health and efficiency. Digital twins stand as vigilant sentinels, monitoring the intricate machinations of engines with unwavering precision. By analyzing myriad data points, they predict potential part failures well before they transpire. This foresight empowers aviation engineers to schedule preemptive maintenance, ensuring the unbroken flight operations of these aerial marvels. Digital twins, for instance, enable engineers to model production procedures, spot any bottlenecks, and optimize assembly line layouts before actual implementation in the car industry. Virtual reality (VR) and augmented reality (AR) can be seamlessly integrated into the manufacturing process with the help of digital twins, providing virtual prototyping, remote help, and immersive training (Ashok et al., 2023). By means of these case studies, we observe digital twins transforming conventional industrial paradigms, promoting flexibility, effectiveness, and adaptability to changing market needs.

Digital Supply Chain Twin

Figure 5 illustrates the complex interrelationships that exist between digital technologies that include supply chain data, different analytics methodologies, and virtual representations of actual goods in supply chain management. Technologies including sensors, cloud computing, blockchain, IoT, and ERP are at the core of this integration. For example, every Tesla sold has a digital representation that reflects its real-world counterpart. That means in 2023, when Tesla's vehicle deliveries grew 38 percent year-over-year, at least 1.8 million corresponding digital twins were created (Business Wire, 2024). Elon Musk is aiming to bring 20 million Teslas per year to production by 2030. Managing digital twins at this scale requires magnificent processing power, data storage, and software infrastructure. Each Tesla digital twin collects and processes real-time data from its corresponding vehicle, including performance metrics, sensor readings, operational conditions, and more (Attaran et al., 2023). The sheer volume of information collected and transmitted to Tesla's data centers requires substantial network bandwidth and infrastructure. From there, complex algorithms and machine learning models extract meaningful insights, which requires the most powerful CPUs and

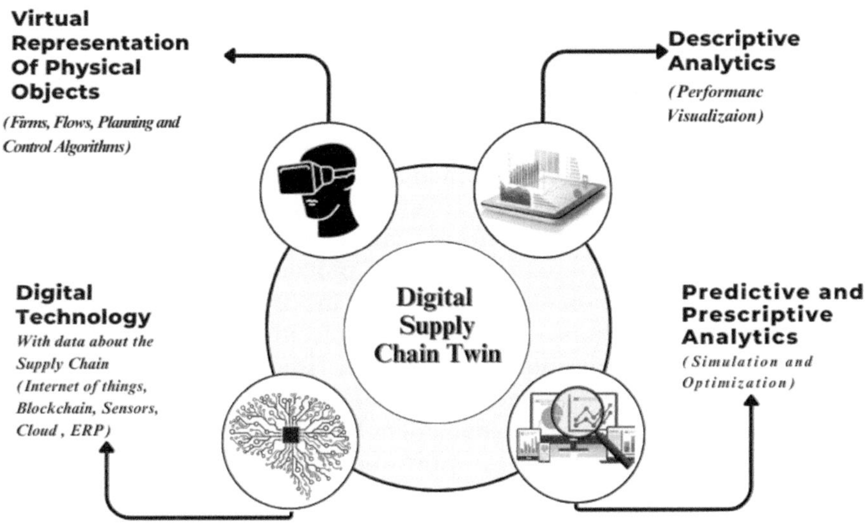

Figure 5: Concept and deisgn of digital Supply chain twin.

GPUs. Data from millions of vehicles, collected over time, must also be stored securely and efficiently, necessitating leading-edge data storage solutions. Tesla uses this real-time data to continuously improve vehicle performance, predict maintenance needs, and offer over-the-air software updates. When these technologies work together, they can virtually represent real-world objects, making supply chain management and monitoring easier. Digital technologies optimize the flow of goods and information between businesses, including distributors, retailers, suppliers, and manufacturers. These technologies, along with planning and control algorithms that make use of data from ERP systems and IoT devices, facilitate the efficient flow of goods and materials across the supply chain.

Smart Cities: Shaping Urban Futures

In smart cities, digital twins are essential to determining how urban development and planning will proceed in the future. City planners generate digital images of whole urban environments by utilizing data from geospatial mapping technology, IoT sensors, and urban sensors. By acting as digital canvases, these twins allow predictive modeling to be used for designing urban infrastructure, energy consumption optimization and traffic control. For example, digital twins help city planners with transportation planning by simulating traffic patterns, pinpointing areas of high congestion, and real-time public transportation routes' optimization (Badshah et al., 2024). Multi-Sensor Fusion Technology's role in sustainable urban governance networks introduces an architectural framework (Pan et al., 2023) underpinning the flow for digital twin platform development aimed at urban condition monitoring. It outlines a use case for the 3D visualization of a smart village in Busan, South Korea, employing a 3D Geospatial platform (Paripooranan et al., 2020) suggests augmented reality (AR)-assisted DT as a pioneering approach towards the future transformation of human-centric industries (Mora et al., 2023) highlights the importance of incorporating innovation management theories into the exploration of smart city transitions, offering novel insights and practical approaches to enhance the governance of smart cities through an innovation management lens. Digital twins are evidence of the potential of technological innovation to encourage more livable, inclusive, and efficient cities as cities throughout the world strive for more adaptability and efficiency.

Aerospace: Optimizing Flight Operations

Digital twins are revolutionizing the design, operation, and maintenance of aircraft in the aerospace sector. Aerospace engineers assemble data from sensor arrays, flight telemetry, and maintenance files to construct digital duplicates of individual aircraft and entire fleets. By acting as virtual cockpits, these twins allow for real-time performance monitoring, predictive maintenance scheduling, and training-related flight simulation (Bano et al., 2022). Engineers can use digital twins, for example, to plan maintenance tasks in advance, predict component failures, and maximize spare part inventories in airplane maintenance. Airbus employs digital twins to continuously monitor and analyze the real-time performance of their A350 XWB aircraft. By gathering data from various sensors and systems, they gain valuable insights into crucial operational factors including fuel consumption, engine efficiency, and overall aircraft performance (Baxter and Srisaeng, 2018). This extensive dataset assists Airbus in identifying areas for optimization, enabling informed decision making based on data analysis to enhance overall performance. Digital twins are also very useful for training, as they allow more accurate flight simulations that improve pilot competency. Pilots can experience and learn from a variety of settings, from normal flights to emergency emergencies, by recreating specific flying scenarios in a virtual environment. This practical instruction was made possible by digital twins, contributing to improved aviation safety and operational effectiveness.

References

Abdulkarem, M., Samsudin, K., Rokhani, F. Z. and Rasid, M. F. (2020). Wireless sensor network for structural health monitoring: A contemporary review of technologies, challenges, and future direction. *Structural Health Monitoring*, 19(3): 693–735. https://doi.org/10.1177/1475921719854528.

Achouch, M., Dimitrova, M., Ziane, K., Sattarpanah Karganroudi, S., Dhouib, R. et al. (2022). On predictive maintenance in Industry 4.0: Overview, models, and challenges. *Applied Sciences*, 12(16): 8081. https://doi.org/10.3390/app12168081.

Ali, Z., Saad, S., Rasheed, K. and Ammad, S. (2024). AI future perspectives and trends in construction. In *AI in Material Science* (pp. 239–261). CRC Press. https://doi.org/10.1201/9781003438489-11.

Ammad, S., Saad, S., Bashir, M. T., Qureshi, A. H., Altaf, M. et al. (2021). Construction accidents via integrating Building Information Modelling (BIM) with emerging digital technologies a review. *2021 3rd International Sustainability and Resilience Conference: Climate Change*. https://doi.org/10.1109/IEEECONF53624.2021.9668066.

Ashok, J., Kumar, N. A., Raj, D. W. P., Ashok, J., Bhushan, A. V. et al. (2023). Design and implementation of smart manufacturing systems through AR for data-driven digital twin system. *SN Computer Science*, 4(5): 580. https://doi.org/10.1007/s42979-023-01956-1.

Attaran, M., Attaran, S. and Celik, B. G. (2023). The impact of digital twins on the evolution of intelligent manufacturing and Industry 4.0. *Advances in Computational Intelligence*, 3(3): 11. https://doi.org/10.1007/s43674-023-00058-y.

Badshah, I., Saad, S., Ammad, S., Rasheed, K., Khan, M. W. et al. (2024). Enhancing sustainable mobility: Lane change strategies for efficient traffic flow at stationary bottlenecks. *2024 ASU International Conference in Emerging Technologies for Sustainability and Intelligent Systems (ICETSIS)*, 1197–1202. https://doi.org/10.1109/ICETSIS61505.2024.10459583.

Baik, J. (Sophia). (2020). Data privacy against innovation or against discrimination? The case of the California Consumer Privacy Act (CCPA). *Telematics and Informatics*, 52: 101431. https://doi.org/10.1016/j.tele.2020.101431.

Bano, D., Michael, J., Rumpe, B., Varga, S., Weske, M. et al. (2022). Process-aware digital twin cockpit synthesis from event logs. *Journal of Computer Languages*, 70: 101121. https://doi.org/10.1016/j.cola.2022.101121.

Barzegar, M., Rajabifard, A., Kalantari, M. and Atazadeh, B. (2021). An IFC-based database schema for mapping BIM data into a 3D spatially enabled land administration database. *International Journal of Digital Earth*, 14(6): 736–765. https://doi.org/10.1080/17538947.2021.1875062.

Baxter, G. and Srisaeng, P. (2018). The strategic deployment of the Airbus A350-900XWB aircraft in a full-service network carrier route network: The case of Singapore airlines. *Infrastructures*, 3(3): 25. https://doi.org/10.3390/infrastructures3030025.

Borgogno, O. and Colangelo, G. (2019). Data sharing and interoperability: Fostering innovation and competition through APIs. *Computer Law & Security Review*, 35(5): 105314. https://doi.org/10.1016/j.clsr.2019.03.008.

Brown, K. M., Hambidge, C. H. and Brownett, J. M. (2016). Progress in operational flood mapping using satellite synthetic aperture radar (SAR) and airborne light detection and ranging (LiDAR) data. *Progress in Physical Geography: Earth and Environment*, 40(2): 196–214. https://doi.org/10.1177/0309133316633570.

Business Wire. (2024, January 2). *Tesla Vehicle Production & Deliveries and Date for Financial Results & Webcast for Fourth Quarter 2023*. https://ir.tesla.com/press-release/tesla-vehicle-production-deliveries-and-date-financial-results-webcast-fourth-quarter-2023.

Choi, W. J., An, J. K., Woo, J. J. and Kwak, H. Y. (2022). Comparison of diagnostic performance in mammography assessment: Radiologist with reference to clinical information versus standalone artificial intelligence detection. *Diagnostics*, 13(1): 117. https://doi.org/10.3390/diagnostics13010117.

Cooper, J. P., Jackson, S., Kamojjala, S., Owens, G., Szana, K. et al. (2022). Demystifying Digital Twins: Definitions, applications, and benefits. *Journal AWWA*, 114(5): 58–65. https://doi.org/10.1002/awwa.1922.

Costa, D. G., Bittencourt, J. C. N., Oliveira, F., Peixoto, J. P. J., Jesus, T. C. et al. (2024). Achieving sustainable smart cities through geospatial data-driven approaches. *Sustainability*, 16(2): 640. https://doi.org/10.3390/su16020640.

de Streel, A., Kraemer, J. and Senellart, P. (2021). Making data portability more effective for the digital economy. *SSRN Electronic Journal, June*. https://doi.org/10.2139/ssrn.3866495.

Fan, C., Zhang, C., Yahja, A. and Mostafavi, A. (2021). Disaster City Digital Twin: A vision for integrating artificial and human intelligence for disaster management. *International Journal of Information Management*, 56: 102049. https://doi.org/10.1016/j.ijinfomgt.2019.102049.

Haris, M., Saad, S., Ammad, S. and Rasheed, K. (2024). AI in fabrication and construction. In *AI in Material Science* (pp. 169–192). CRC Press. https://doi.org/10.1201/9781003438489-8.

Haris, M., Saad, S., Ammad, S., Rasheed, K., Aman, H. et al. (2024). Integrating sustainable practices for effective solid waste management in peshawar: Overcoming challenges and promoting environmental. *2024 ASU International Conference in Emerging Technologies for Sustainability and Intelligent Systems (ICETSIS)*, 1–7. https://doi.org/10.1109/ICETSIS61505.2024.10459445.

Javaid, M., Haleem, A. and Suman, R. (2023). Digital Twin applications toward Industry 4.0: A review. *Cognitive Robotics*, 3: 71–92. https://doi.org/10.1016/j.cogr.2023.04.003.

Jeddoub, I., Nys, G. -A., Hajji, R. and Billen, R. (2023). Digital Twins for cities: Analyzing the gap between concepts and current implementations with a specific focus on data integration. *International Journal of Applied Earth Observation and Geoinformation*, 122: 103440. https://doi.org/10.1016/j.jag.2023.103440.

Khan, M. I., Khan, N., Saad, S. and Khan, M. W. (2024). Artificial Intelligence in construction materials. In *AI in Material Science* (pp. 47–64). CRC Press. https://doi.org/10.1201/9781003438489-3.

Khan, N., Yaqoob, I., Hashem, I. A. T., Inayat, Z., Mahmoud Ali et al. (2014). Big Data: Survey, technologies, opportunities, and challenges. *The Scientific World Journal*, 2014: 1–18. https://doi.org/10.1155/2014/712826.

Lim, K. Y. H., Zheng, P. and Chen, C. -H. (2020). A state-of-the-art survey of Digital Twin: techniques, engineering product lifecycle management and business innovation perspectives. *Journal of Intelligent Manufacturing*, 31(6): 1313–1337. https://doi.org/10.1007/s10845-019-01512-w.

McKinney, S. M., Sieniek, M., Godbole, V., Godwin, J., Antropova, N. et al. (2020). International evaluation of an AI system for breast cancer screening. *Nature*, 577(7788): 89–94. https://doi.org/10.1038/s41586-019-1799-6.

Mora, L., Gerli, P., Ardito, L. and Messeni Petruzzelli, A. (2023). Smart city governance from an innovation management perspective: Theoretical framing, review of current practices, and future research agenda. *Technovation*, 123: 102717. https://doi.org/10.1016/j.technovation.2023.102717.

Moreno-Gomez, A., Perez-Ramirez, C. A., Dominguez-Gonzalez, A., Valtierra-Rodriguez, M., Chavez-Alegria, O. et al. (2018). Sensors used in structural health monitoring. *Archives of Computational Methods in Engineering*, 25(4): 901–918. https://doi.org/10.1007/s11831-017-9217-4.

Naeem, U. H., Bilal, M., Syed, F., Rasheed, K., Saad, S. et al. (2022). A multilayer encryption model to protect healthcare data in cloud environment. *2022 International Conference on Data Analytics for Business and Industry, ICDABI 2022*. https://doi.org/10.1109/ICDABI56818.2022.10041708.

Naqvi, M., Gilani, S. Q., Syed, T., Marques, O., Kim, H. -C. et al. (2023). Skin cancer detection using deep learning—A review. *Diagnostics*, 13(11): 1911. https://doi.org/10.3390/diagnostics13111911.

Oesterreich, T. D. and Teuteberg, F. (2016). Understanding the implications of digitisation and automation in the context of Industry 4.0: A triangulation approach and elements of a research agenda for the construction industry. *Computers in Industry*, 83: 121–139. https://doi.org/10.1016/j.compind.2016.09.006.

Otto, B. and Jarke, M. (2019). Designing a multi-sided data platform: Findings from the International data spaces case. *Electronic Markets*, 29(4): 561–580. https://doi.org/10.1007/s12525-019-00362-x.

Pan, Z., Zhang, Z., Meng, Z. and Wang, Y. (2023). A novel fault classification feature extraction method for rolling bearing based on multi-sensor fusion technology and EB-1D-TP encoding algorithm. *ISA Transactions*, 142: 427–444. https://doi.org/10.1016/j.isatra.2023.07.015.

Paripooranan, C. S., Abishek, R., Vivek, D. C. and Karthik, S. (2020). An implementation of AR enabled Digital Twins for 3-D printing. *2020 IEEE International Symposium on Smart Electronic Systems (ISES) (Formerly INiS)*, 155–160. https://doi.org/10.1109/iSES50453.2020.00043.

Piasecki, S. (2023). Expert perspectives on GDPR compliance in the context of smart homes and vulnerable persons. *Information & Communications Technology Law*, 32(3): 385–417. https://doi.org/10.1080/13600834.2023.2231326.

Rasheed, K., Ammad, S., Said, A. Y., Balbehaith, M., Oad, V. K. et al. (2022). Application of smart IoT technology in project management scenarios. *2022 International Conference on Data Analytics for Business and Industry, ICDABI 2022*. https://doi.org/10.1109/ICDABI56818.2022.10041674.

Rasheed, K., Saad, S., Ammad, S. and Badshah, I. (2023). IoT enhanced solar device management: Integrating remote surveillance and energy efficiency for sustainable operation. *2023 TRON Symposium (TRONSHOW)*, 1–6.

Rasheed, K., Saad, S., Ammad, S. and Bashir, M. T. (2024). Industry 4.0 and construction. In *AI in Material Science* (pp. 65–87). CRC Press. https://doi.org/10.1201/9781003438489-4.

Rasheed, K., Saad, S., Ammad, S. and Bilal, M. (2023). Integrating sustainability management and lean practices for enhanced supply chain performance: Exploring the role of process optimization in SMEs. *Engineering Proceedings*, 56(1). https://doi.org/10.3390/ASEC2023-16370.

Rasheed, K., Waqas, M., Saad, S., Zaland, A., Ali, Z. et al. (2024). Ethical and societal implications of AI in construction. In *AI in Material Science* (pp. 216–238). CRC Press. https://doi.org/10.1201/9781003438489-10.

Rasheed, K., Zaland, A., Saad, S., Ammad, S., Rostami, A. et al. (2024). History of AI. In *AI in Material Science* (pp. 15–46). CRC Press. https://doi.org/10.1201/9781003438489-2.

Rodriguez, B. X., Huang, M., Lee, H. W., Simonen, K., Ditto, J. et al. (2020). Mechanical, electrical, plumbing and tenant improvements over the building lifetime: Estimating material quantities and embodied carbon for climate change mitigation. *Energy and Buildings*, 226: 110324. https://doi.org/10.1016/j.enbuild.2020.110324.

Saad, S., Alaloul, W. S., Ammad, S., Qureshi, A. H., Altaf, M. et al. (2020). Design phase carbon emission prediction using a visual programming technique. *2020 2nd International Sustainability and Resilience Conference: Technology and Innovation in Building Designs*. https://doi.org/10.1109/IEEECONF51154.2020.9319978.

Saad, S., Ammad, S., Rasheed, K., Balbehaith, M., Qureshi, A. H. et al. (2023). Enhancing construction management in restoration through BIM-assisted photogrammetric 3D model accuracy assessment. *2023 International Conference on Sustaining Heritage: Innovative and Digital Approaches (ICSH)*, 100–103. https://doi.org/10.1109/ICSH57060.2023.10482845.

Saad, S., Haris, M., Ammad, S. and Rasheed, K. (2024). AI-assisted building design. In *AI in Material Science* (pp. 143–168). CRC Press. https://doi.org/10.1201/9781003438489-7.

Saad, S., Rasheed, K., Ammad, S., Ali, Z., Zaland, A. et al. (2024). The conclusive role of AI in Industry 4.0 construction naterials. In *AI in Material Science* (pp. 262–270). CRC Press. https://doi.org/10.1201/9781003438489-12.

Saad, S., Rasheed, K., Ammad, S., Waqas Khan, M., Zaland, A. et al. (2024). Is AI the architect of tomorrow's materials in the age of Industry 4.0? In *AI in Material Science* (pp. 1–14). CRC Press. https://doi.org/10.1201/9781003438489-1.

Saad, S., Zaland, A., Ammad, S. and Rasheed, K. (2024). Robotics in construction. In *AI in Material Science* (pp. 115–142). CRC Press. https://doi.org/10.1201/9781003438489-6.

Shahzad, L., Rasheed, K., Ahmad, I., Haider, S. W., Saad, S. et al. (2021). Cost effective accident prevention system for vehicles. *2021 International Conference on Decision Aid Sciences and Application, DASA 2021*. https://doi.org/10.1109/DASA53625.2021.9682266.

Song, Y., Mao, F. and Liu, Q. (2019). Human comfort in indoor environment: A review on assessment criteria, data collection and data analysis methods. *IEEE Access*, 7: 119774–119786. https://doi.org/10.1109/ACCESS.2019.2937320.

Tang, S., Shelden, D. R., Eastman, C. M., Pishdad-Bozorgi, P., Gao, X. et al. (2019). A review of building information modeling (BIM) and the internet of things (IoT) devices integration: Present status and future trends. *Automation in Construction*, 101: 127–139. https://doi.org/10.1016/j.autcon.2019.01.020.

Valinejadshoubi, M., Moselhi, O. and Bagchi, A. (2022). Integrating BIM into sensor-based facilities management operations. *Journal of Facilities Management*, 20(3): 385–400. https://doi.org/10.1108/JFM-08-2020-0055.

Vishwamitra, N., Li, Y., Hu, H., Caine, K., Cheng, L. et al. (2022). Towards automated content-based photo privacy control in user-centered social networks. *Proceedings of the Twelfth ACM Conference on Data and Application Security and Privacy*, 65–76. https://doi.org/10.1145/3508398.3511517.

Waqas Khan, M., Saad, S., Ammad, S., Rasheed, K., Jamal, Q. et al. (2024). Smart infrastructure and AI. In *AI in Material Science* (pp. 193–215). CRC Press. https://doi.org/10.1201/9781003438489-9.

Zaland, A., Saad, S., Ahmad, W., Ammad, S., Khan, M. W. et al. (2024). Enhancing mechanical properties of concrete by adding polypropylene fibers. *2024 ASU International Conference in Emerging Technologies for Sustainability and Intelligent Systems (ICETSIS)*, 771–778. https://doi.org/10.1109/ICETSIS61505.2024.10459663.

Zaland, A., Saad, S., Rasheed, K. and Ammad, S. (2024). Automation in construction industry. In *AI in Material Science* (pp. 88–114). CRC Press. https://doi.org/10.1201/9781003438489-5.

Zanke, A., Weber, T., Dornheim, P. and Engel, M. (2024). Assessing information security culture: A mixed-methods approach to navigating challenges in international corporate IT departments. *Computers & Security*, 144: 103938. https://doi.org/10.1016/j.cose.2024.103938.

Zhang, J. and Lin, X. (2017). Advances in fusion of optical imagery and LiDAR point cloud applied to photogrammetry and remote sensing. *International Journal of Image and Data Fusion*, 8(1): 1–31. https://doi.org/10.1080/19479832.2016.1160960.

Chapter 4

Robotics and Automation
Tools and Technologies

Zawar Ali,[1] *Kumeel Rasheed,*[1,*] *Syed Saad,*[2]
Syed Ammad[3] and *Luqman Shahzad*[4]

Introduction

In the dynamic realm of construction, robotics and automation have emerged as transformative forces redefining established benchmars for efficiency and novelty. This chapter embarks on a comprehensive exploration of robotics and automation: tools and techniques, tracing their origins, examining their current applications, and envisioning their future potential. Robotics and automation emerged from an insatiable hunger for perfection, productivity, a conceptual rethinking of everyday traditions as carried out in different fields. Originally the pillars of manufacturing, these technologies have undergone an incredible transformation journey that has seen them break free from their original application limitations. The integration of robotics and automation is an endeavor that crosses sectors to leave a definitive imprint upon the studies on technology. Farming and construction are two sectors where robotics and automation represent a departure from traditional labor-intensive methodologies. These technologies no longer consign construction to

[1] Department of Civil Engineering, CECOS University of IT and Emerging Sciences, Pakistan.
[2] Science, Engineering & Manufacturing Centre Sandwell College West Bromwich, United Kingdom.
[3] School of Engineering and Built Environment, Griffith University, Gold Coast, Australia.
[4] School of Computing Sciences, Pak-Austria Fachhochschule Institute of Applied Sciences and Technology, Pakistan.
* Corresponding author: kumeelrasheed@cecosian.edu.pk

perform dull tasks but manage a paradigm shift, positioning the trade in front of innovation and efficiency.

Within the current environment, robotics and automation have effortlessly become one of many factors shaping solutions in the construction industry. With the arrival of digital twins, which replicate physical assets in a virtual environment, complexity makes its way into decision-making procedures. This integration is not simply a vehicle for efficiency improvements; it is an entirely new model of how construction projects are envisioned, considered and implemented. Picture a construction site where digital twins work in alignment with their physical twins, providing real-time information transfer, risk mitigation and improved collaboration. This is the physical reality drawn by robotics and automation, a canvas where each brush stroke symbolizes the ever-changing nature of construction technology. As the chapter explores what is inside the robotics and automation toolbox, it discusses many different technologies driving change. Moving beyond complex terminology, the chapter engages in discussion, elucidating these instruments in a way that is clear and understandable. At its core, this technological synergy depends on sensors and actuators, serving as the neurological and muscular parts of the operation. Once seen only as machine pieces, these components have evolved into designers of precision, fostering smooth interaction between the physical and digital spaces. Sophisticated programming languages and machine-gaining algorithms, frequently hidden in complex language, become clear tools in this explanation. Standing at the junction of the future, the path of robotics and automation within construction appears both promising and transformative. The integration of digital doubles with robotics is set to redefine the parameters of design, construction, and operation (Iturralde et al., 2022). This is not merely a technological evolution; it is an ongoing revolution with far-reaching implications.

- This chapter examines the complex relationship between robotics and automation, revealing their symbiotic nature and presenting real-world applications.
- The exploration bridges the gap between theory and practical implementation, ensuring every concept is not just understood but applied.
- Deeper exploration reveals robotics and automation's convergence as a vital force collected to redefine efficiency and innovation norms.
- This chapter sets the stage for an in-depth look at robotics and automation tools/techniques, unraveling roots, current applications and promising trajectories.

- The genesis of robotics and automation lies in pursuing precision, efficiency and reimagining conventional practices across industries.
- Originating as manufacturing experts, these technologies have expanded beyond initial applications through transformative journeys.
- The union of robotics and automation is an evolutionary sequence permeating diverse sectors with permanent technological impacts.

Robotics Hardware and Software Integration

The cooperative integration of robotics hardware and software components forms the foundation enabling automation across industries. Within construction, robotic technologies continue advancing with improved sensors, effectors, mobility, and artificial intelligence (AI). However, maximizing their benefits requires comprehending the technologically consolidated physical forms with virtual control architectures. At their core, robots consist of gifted mechanical structures blending with intelligent software brains. End effectors allow physical interaction, including arms, hands, welders, and tools configured for tasks. An assortment of external sensors provides environmental awareness, from cameras capturing visual data to encoders measuring joint positions. Proprioceptive sensors also enhance self-awareness. Actuators powered by electric motors drive movement through linkages, mirroring muscle functions with joints (Leder and Menges, 2023). These hardware pieces seamlessly interface through programmable control systems. Low-level machine code focuses on basic behaviors while higher levels introduce sequencing, logic and decision making. Distributed control architectures split processing across modular nodes for enhanced responsiveness. Programmable logic controllers rapidly organize input-output synchronization between physical and virtual components (Haris et al., 2024).

User-friendly programming environments empower non-experts to design multi-step routines visually versus writing code. Control systems integrate real-time data streams to build a contextual understanding. Feedback control algorithms enable precise force/motion regulation through closed-loop proportional-integral-derivative control strategies. Wireless connectivity facilitates remote supervision via networks. Path planning and autonomous reasoning enhance mobility. Spatial mapping locates obstacles/surfaces for navigation. Machine vision perceives imagery to comprehend contexts. Machine learning algorithms identify patterns to tune motions for unfamiliar tasks through trial-and-error interactions. Force/tactile sensors provide touch sensitivity. Together, these hardware and software innovations endow robots with versatile anthropic abilities. Their tight coordination underlies sophisticated automated capabilities across domains. Within construction, strategic integration can optimize workflows with precision, safety, and productivity far surpassing manual counterparts (Saad et al., 2024). Comprehending cyber-physical coupling empowers builders to realize robotic potential through informed implementation and management. Ultimately, it is the next technological transformation across the industry (Zhang et al., 2023).

Robot Architecture

The architectural design of a robot forms the physical basis enabling it to operate within the real world. Considered at both microscopic and macroscopic levels, a robot's architecture carefully organizes constituent components into an integrative framework balancing function with form. At the core of any robotic system lies its embodied hardware arranged to achieve targeted objectives. End effectors represent the biological analogy of hands, directly interfacing to manipulate environments (Leder and Menges, 2023). Common varieties include articulated arms consisting of linked rotational segments extending workspaces. Anthropomorphic designs mimic human musculature and range of motion for intuitive operation. Wristed platforms flexibly orient end-tip payloads. Parallel grippers open/close to grasp objects while vacuum-based styles securely hold irregular shapes. End effectors represent the biological analogy of hands, directly interfacing to manipulate environments. Common varieties include articulated arms consisting of linked rotational segments extending workspaces. Anthropomorphic designs mimic human musculature and range of motion for intuitive operation. Wristed platforms flexibly orient end-tip payloads (Huang et al., 2023). Parallel grippers open/close to grasp objects while vacuum-based styles securely hold irregular shapes. Specialized end effectors include cutting, welding and sealing tools adjusted for precision tasks. Spray guns coat surfaces while lab robots leverage pipettes, clamps and tweezers for biological applications. Cameras, scopes or sensors may substitute arm structures, collecting visual data. Multi-fingered hands and versatile palms facilitate skillful manipulation of parts for assembly. The chosen end effector dictates maximum functional capabilities.

Sensors gather external feedback enabling contextual awareness. Vision sensors capture color images supporting computer analyses. Time-of-flight cameras and Light Detection and Ranging (LiDAR) rapidly build 3D maps for navigation. Force-torque sensors measure contact mechanics, while proximity, strain, and tactile sensors detect interactions. Environmental sensors monitor factors such as temperature, humidity, smoke, and chemicals, while joint encoder readings provide precise information about configuration. Distributed sensor arrangements optimize coverage and minimize noise interference, with sensor selection depending on the required perceptual abilities. With processing power increasing exponentially, sensor fusion methodologies intelligently correlate multimodal inputs (Waqas Khan et al., 2024). Combined with machine learning, this enhances real-time decision-making. Electric motors actuate joints in rotational or linear fashions through transmission systems. Harmonic drives, leadscrews and belt mechanisms translate torques to smooth, accurate and repeatable movements (Shayesteh et al., 2023). On-board motor controllers adjust speed and position through closed-loop feedback. Lithium-polymer battery modules or wired power ports supply constant voltage. Structural platforms support payloads during dynamic tasks, with material selections considering strength, weight, heat dissipation, and longevity (Zaland et al., 2024). For outdoor robots, durable seals and housing are essential. As technology advances, lighter stronger alloys and alternative materials will push boundaries for heavier payloads and expanded mobility. Overall, a robot's architecture emerges from the complex optimization of its embodied components working in harmony. Balancing flexibility, sensing and actuation enables versatile applications across industries through intelligent engineering. As shown in Figure 1 the bar chart illustrates the distribution of hardware and software components in a robot's architecture, providing an overview of the efficient optimization and balance between different elements.

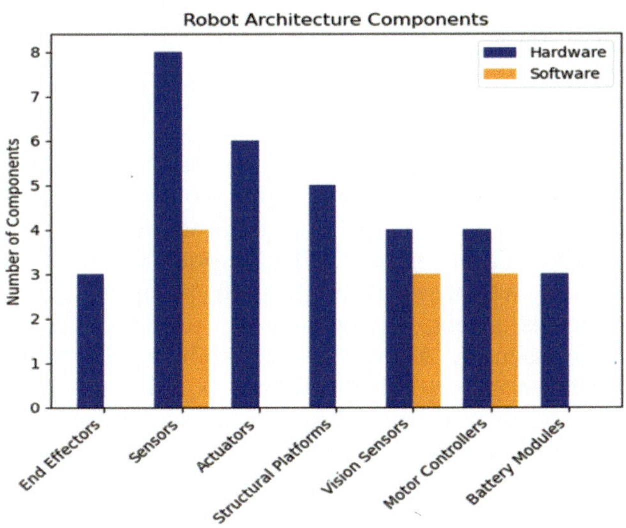

Figure 1: Distribution of components in a robot's architecture.

Control Systems

At the core of any robot lies its control system, functioning analogously to a nervous system. It integrates diverse hardware and software components regulating behavior in synchrony with real-time feedback. The lowest level machine code commands basic motor functions and sensor inputs/outputs. Progressing layers introduce higher-order semantics for sequencing tasks, conditional logic, and complex decision-making based on environmental understanding (Shayesteh et al., 2023). Distributed control architectures partition processing across modular nodes communicating wirelessly, enabling real-time responsiveness beyond single controller capabilities. Programming languages abstract hardware interactions for portability between diverse robotic platforms. Real-time operating systems ensure deterministic execution without interruptions disrupting smooth operations critical for precise automation (Zaland et al., 2024). Programmable logic controllers rapidly iterate input/output cycles maintaining synchronization between digital control and analog physical processes (Oke et al., 2023). Field programmable gate arrays provide reconfigurable hardware acceleration for computer vision and other perception intensive algorithms. Control systems thus optimize distributed embedded computing architectures matching robotic complexities.

Natural visual interfaces empower non-expert users through graphical workflows avoiding text-based coding. Navigation between common nodes fosters rapid prototyping of diverse routines. Playback validates operation as intended before deployment and version control facilitates collaborative programming extending individual knowledge. Offline simulation verifies safe, realistic behavioral responses to uncertainties, integrating real-time sensor data streams to build situational understandings. Proprioceptive readings continuously map robot self-geometry for repeatability, while control loops based on PID algorithms regulate forceful positional accuracy in delicate interactions with environments (Tapia et al., 2020). Computer vision perceives visual inputs, directing responses through path planning algorithms that navigate obstacles safely. Tactile sensing protects against excessive contacts during assembly. Autonomous capabilities emerge from machine learning, which recognizes patterns within experiences. Motion libraries optimize grasping of unknown objects, whereas reinforcement learning fine-tunes policies for handling new, unforeseen scenarios. Constrained optimization routines consider dynamics safely executing trajectories. Distributed computing architectures support ever more complex reasoning with increasing onboard processing and cloud offloading (Naeem et al., 2022). Overall, the tight integration of a robotic control system's constituent hardware and software components empowers truly intelligent behavior exceeding human physical abilities through dynamic feedback interactions within dynamic environments (Iturralde et al., 2022). Their realization underlies precision automated applications across industries revolutionizing how tasks are performed.

Revolutionizing Automation: Equipment, Machinery, and Beyond

The ongoing transformation initiated by robotics and automation is redefining industries globally at an unprecedented pace. Originally concentrated primarily in manufacturing facilities, the revolutionary impacts of these technologies now permeate far beyond narrow industrial confines into nearly all realms of modern civilization (Khan et al., 2024). Advanced robotic and automated equipment and machinery epitomize the most tangible and visible fronts of this ongoing evolutionary process. However, the changes unleashed extend far deeper than just novel industrial applications might suggest. Powerful transformative ripples generated in their wake likewise influence surrounding societal layers in detail and complex ways that will play out over the coming decades. Business models, employment practices, consumer habits and many cultural norms face disruption as smart machines become ever more ingrained across our daily lives (Kaur et al., 2023). Within manufacturing, industrial robots have rapidly supplanted humans in performing repetitive, hazardous or ergonomically challenging tasks with levels of accuracy, consistency, reliability and throughput impossible for flesh and blood workers (Saad et al., 2024). Collaborative robots now work safely alongside employees, combining strengths to optimize production. Mobile robotic platforms autonomously transport raw materials and finished goods, freeing plant personnel for higher-level roles. Manufacturing lines can easily reconfigure through modular, reprogrammable automated equipment to flexibly accommodate design changes or mass customization. Additive manufacturing techniques like 3D printing have empowered the on-demand fabrication of highly customized components without costly tooling. Micro-factories leveraging such technologies are enabling localized, smaller-batch production near end users and markets. Detailed digital inventories and real-time monitoring solutions optimize just-in-time delivery while minimizing warehousing. The new paradigms of robotics and automation have unlocked entirely undreamed-of optimization avenues such as these for core industrial operations. Construction has similarly witnessed the rapid infiltration of automation. Intelligent machines now simulate complex building information models, validating logistics and construction sequencing while identifying issues during the design phase (Baduge et al., 2022). Robotic assistants perform dangerous jobs like welding at heights, reducing risks for human workers. Automated quality control scans completed work against digital CAD models, identifying deviations early to minimize reworks. Additionally, drones rapidly map and monitor expansive worksites. Prefabricated components cut at factories are assembled more efficiently onsite through robotic methods.

Beyond conventional industrial spheres, flexible new applications for robotics and automation continue emerging across diverse domains. Robotic systems augment human capabilities for safer material handling in warehouses and distribution centers (Zaland et al., 2024). Autonomous trucks and vessels transport shipping containers across countries and oceans with precision navigation. Agricultural robots weed, fertilize, and monitor vast fields with customized care. Assistive robots aid the disabled and elderly with independence. Autonomous floor cleaning machines maintain buildings efficiently (Halder et al., 2023). Service robots interact directly with customers through reception, food delivery, or telepresence. The only limits lay in our imagination for creative deployment avenues. Naturally, these deeply transformative technological shifts have prompted profound business model adaptations. Mass customization unlocked by on-demand manufacturing, 3D printing and similar flexibilities allows niche markets formerly deemed unfeasible. Digitally integrated value chains minimize overheads through predictive analytics, condition-based maintenance, centralized remote monitoring and more. Virtual inventories and just-in-time logistics substantially reduce warehousing footprints, while new verticals around robotics have blossomed, generating further socioeconomic impacts.

For employers and workers, roles now increasingly attract towards conceptual, care-based, customer-facing and managerial functions as manual burdens are lifted. A reskilling challenge has emerged for personnel to adapt, with younger generations entering workforce landscapes emphasizing science, technology, engineering and mathematics (STEM) perception, computational thinking, creativity and lifelong learning compared to prior eras. Coexistence models also promise new vistas, combining complementary human and machine skillsets for mutual benefit. In closing, while industrial revolutions of the past dominated certain sectors, robotics and automation show the potential to transform civilization at foundations across geographic and industry boundaries (van der Heijden, 2023). Their societal impact is still evolving, but the likely winners will be those who proactively shape this new paradigm rather than resist its progress. Visionaries who optimize human-machine symbiosis hold the greatest hope for a prosperous, fulfilling future.

Industrial Robotics

At the cutting edge of the industrial revolution driven by robotics and automation are the machines that have come to dominate factories around the world - industrial robots. Designed for operation on the shop floor, these systems excel at tasks demanding attributes beyond human physical or cognitive abilities such as lifting or maneuvering heavy payloads, performing repetitive motions, withstanding hazardous environmental conditions and achieving microscopic precision (Jiang et al., 2023). On traditional manufacturing assembly lines, industrial robots tirelessly perform welding, painting, packaging and similar operations. Their durability and precision allow for the production of thousands of identical units per hour with perfect consistency far exceeding any person. Force-feedback and collision detection safely stop motion upon unexpected contact preventing injuries. Meanwhile, their integrated vision systems flawlessly discern and manipulate parts, enabling flexible production and accommodating frequent design changes through prompt reprogramming. More recently, collaborative robots or "cobots" have entered the scene. Capable of operating safely alongside human coworkers without heavy safety cages, collaborating efficiently on joint tasks (Dui et al., 2023). Cobots intuitive programming methods empower non-experts to train new motions through online or offline teaching, speeding redeployment between diverse applications.

Mobile robotic platforms now augment work throughout factories, autonomously transporting raw materials, work-in-progress items and finished goods precisely as required to optimize material flows (Saad et al., 2024). Remote operation enables transporting payloads too massive or hazardous for human handling with meticulous navigation. Robotic systems worn by employees lift or transfer heavy components, reducing strains on their bodies to support safer, more sustainable logistics. Advanced vision-based quality control systems scrutinize products, detecting even microscopic variances in properties such as dimensions, finishes or part orientations. Machine learning algorithms enhance discerning flaws over time. Additive manufacturing now fabricates mass customized or personalized components on demand through 3D printing or similar techniques. By eliminating the need for metallic tooling, factories can reduce changeover times and swiftly adapt to changing demands.

Micro factories locate modular production cells in close proximity to end users through contract manufacturers. From aviation components to medical devices, they manufacture low-volume specialized products on schedule without outsourcing overseas (Dui et al., 2023). Adaptable facilities minimize downtime by reallocating machines between roles.

Remote monitoring ensures consistent quality worldwide through preventative maintenance predicted by real-time usage metrics (Rasheed et al., 2024). As technologies progress, new frontiers will emerge. Nevertheless, industrial robotics has already revolutionized manufacturing by implementing coordinated automation, resulting in significantly greater precision, output, safety, and flexibility compared to traditional methods. They optimize processes across industries through augmenting rather than replacing human capabilities (Halder et al., 2023). In Figure 2, the industrial landscape unfolds, showcasing the key players: humans, industrial robots, and collaborative robots. Each entity possesses unique attributes, from the human's strength to the collaborative robot's adaptability. The true innovation lies in the synergy, likened to an orchestra, where humans and cobots work together in harmony. Attributes such as durability, precision, and safety serve as the foundational elements, while human skills and cost-effectiveness contribute to the melody. This collaboration represents a significant

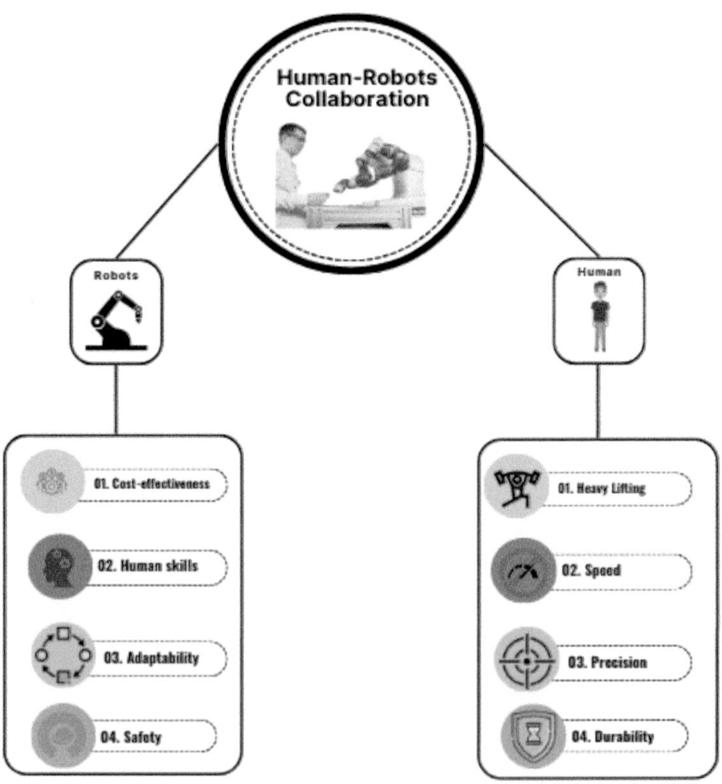

Figure 2: Humans and robots crafting a harmonious production integration.

partnership, blending steel and human effort, marking a new era in production dynamics.

Construction Automation

The construction industry has entered the early stages of a robotic revolution with immense potential to enhance safety, quality and productivity. Mobile robotics now mobilize heavy machinery with centimeter precision through integrated global navigation satellite systems (GNSS) and sensors. Bulldozers precisely grade sites and pavers lay uniform asphalt and concrete according to digital blueprints. Robotic total stations autonomously survey areas, capturing 4D progress scans for project planners. Drones rapidly map extensive worksites, identifying issues undetectable from ground level like drainage flaws or structural cracks in buildings. Thermal and infrared cameras detect anomalies before they are visible to the naked eye. Computer vision processes imagery, automatically flagging deviations from plans (Oke et al., 2023). Robotic assistants now safely perform hazardous tasks otherwise too dangerous for humans, like welding at heights on scaffolds or in confined underground spaces. Robotic systems worn by workers counterbalance or assist in lifting increasingly heavy and bulky rebar, wall panels and beams for safer manual labor.

Intelligent modeling software simulates complex building information models (BIM), automatically validating construction sequencing, material requirements, logistics and more to catch conflicts early in design (Oke et al., 2023). Automated quality control robots equipped with laser scan works against as-planned CAD models, precisely identifying measurement errors or faulty installations for timely remedy. Robots controlled through augmented reality interfaces then precisely install intricate piping, ductwork, cabling and framework demanding perfection (Ali et al., 2024). Prefabricated components like walls, floors, facade panels and trusses assembled offsite by collaborative robots shorten onsite timelines. Self-driving vehicles autonomously deliver supplies precisely to crews as needed, eliminating traffic congestion and accidents (Zhao et al., 2023). Drones inspect structures for corrosion or maintenance issues, detecting problems before catastrophic failures occur through computer vision and thermal anomaly detection. Overall, construction robotics empower higher quality vertical builds with unprecedented safety, efficiency and flexibility. As digital industrialization advances, new frontiers will continue to emerge. There is speculation that exoskeletons could eventually be developed to lift entire floors into place. 3D printing will be used to fabricate complex architectural features on demand, while swarms of collaborative drones and robots could autonomously erect

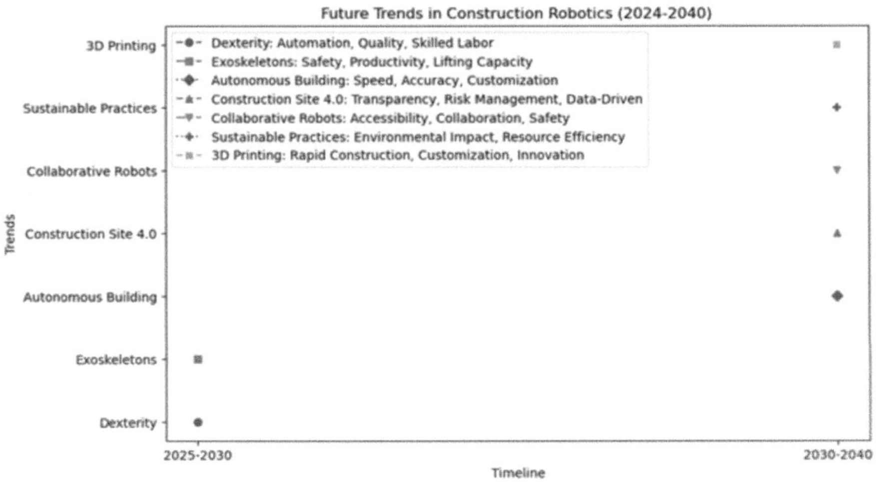

Figure 3: Evolution of construction robotics.

buildings from raw materials. Real-time wireless diagnostics can prevent life-threatening accidents through predictive maintenance (Rasheed et al., 2024). Ultimately, people and machines will partner symbiotically to maximize creativity and productivity through shared strengths. The built environment of tomorrow will be one shaped as much by code as concrete through revolutionary advances empowering the industry today. Figure 3 illustrates future trends in construction robotics (2024–2040). Each trend is represented by a unique marker and line style, aiding visual comparison. Key impacts and timelines are detailed for enhanced clarity.

Human-Robot Collaboration in Construction

The global construction sector stands upon the threshold of a revolutionary digital transformation as pioneering integration of human expertise and robotic assistance takes root. Whereas manufacturing spearheaded automation's arrival through numerically controlled machinery decades earlier, the complexity realities of vast and constantly evolving outdoor job sites long weathered construction against widespread robotic infiltration. However, recent breakthroughs now empower unprecedented synergistic collaboration between human ingenuity and cybernetic productivity (Saad et al., 2024). On worksites, autonomous mobile robotic platforms equipped with advanced sensors autonomously follow crews throughout shifts, assisting with material handling, preliminary welding, and inspection scans. Exoskeletons augment workers' lifting and fastening tasks, detecting unsafe strains through biometrics to prevent injuries from repetitive motions or heavy loads (Zhang et al., 2023). Robotic arms equipped with force-feedback and micro-precision functionality smoothly install heavy rebar cages, pipe networks and cabling with centimeter level accuracy. Computer vision algorithms processed through on-board or cloud-based systems rapidly analyze imagery captured by surveillance drones, flagging deviations early for timely remedy and optimized planning. Self-driving vehicles equipped with real-time localization technologies precisely deliver supplies to crews via geofencing as dictated by integrated schedules (Sun et al., 2023). Alongside physical robots, desktop technologies similarly enhance project design and management. Sophisticated BIM software simulates entire constructions digitally from their earliest phases, optimizing workflows, catching issues proactively through structural and operational analyses (Ammad et al., 2021). Automated laser scanning systems precisely discern completed works against digital engineering plans upon project milestones, outlining discrepancies needing addressing and approving phases for continuation. Augmented and virtual reality interfaces guide crews, exoskeletal workers and robots through intuitive visual controls. Advanced remote collaboration solutions transmit expertise globally, facilitating oversight of expansive projects. Meanwhile, evolved programming approaches including hand-free teach pendants simplify rapidly retraining robots between evolving tasks. As automation continues infiltrating construction, the new human-robot dynamic empowers elevating output quality and workforce wellbeing to unprecedented heights (Rasheed et al., 2021). Off-site within fabrication facilities, collaborative industrial robots accelerate production of modular prefabricated construction components that were traditionally manufactured through intensive manual labor. It precisely cuts wall panels, trusses and rebar cages to size according to BIM data, fasten assemblies, smoothen surfaces, apply protective finishes and conduct quality assurance scans often operating safely in direct proximity to human coworkers.

Job Site Automation

From construction's initiation, worksites have engaged armies of laborers exerting immense physical strain under often hazardous conditions. While technology slowly seeped in through machinery improvements, automated solutions remained confined to niche applications like 3D laser scanning or hazardous welding until recently (Mishler and Chen, 2023). However, revolutionary mobile robotics now empower transforming production through seamless human-machine collaboration. Early on, robots primarily focused on individual automated subtasks such as surveillance or danger mitigation. Static industrial arms were used for welding pipes or performing repetitive motions, which provided marginal improvements in efficiency. Now, versatile mobile platforms equipped with advanced sensors autonomously accompany crews throughout dynamic outdoor environments, assisting diversely. Robotic systems worn by workers counterbalance or power tools and materials, detecting biometrics to prevent strains from lifting or repetitive stresses (Filippi et al., 2023). Robotic manipulator arms smoothly handle rebar cages, pipes, sheets and cabling for framing and infrastructure with micro-precision. Force controlled motions ensure safety alongside people while meticulously meeting engineering tolerances. Drones equipped with 4K cameras rapidly capture site wide photogrammetry scans for planning, scheduling optimization and virtual inspections. Computer vision algorithms processed on-board or via cloud middleware automatically flag deviations from plans requiring remedy (Rasheed et al., 2024). Localization technologies incorporating GNSS, visual odometry and simultaneous localization and mapping (SLAM) dynamically map changing spaces for navigation amid obstacles or adverse conditions. IoT sensors networked across equipment monitor utilization and maintenance needs, maximizing uptime through predictive analytics.

Prefabrication Automation

The construction industry is undergoing profound transformations through intelligent integration of robotic automation within prefabricated product manufacturing. As mobile robotics revolutionize jobsite workflows, parallel revolutions maximize off-site production through collaborative human-machine partnerships (Yevu et al., 2023). While manual precasting formerly imposed intensive labor demands, advanced robots now accelerate output of standardized and customized prefab modular components alongside skilled human artisans. Advanced industrial robotic arms precisely cut intricate wall panels, trusses, beams and other subassemblies to micrometer tolerances seamlessly coordinated with digital BIM (Anane et al., 2022). Integrated vision systems judiciously guide robots in distinguishing adjacent co-workers through real-time machine perception. Within dynamic manufacturing environments, self-localizing robots dynamically maneuver obstacles through continuous kinematic modeling anchored by laser tracker triangulation or visual SLAM. Advanced six-axis load cells and torque transducers continually monitor robot poses, velocities and contact forces, automatically pausing tasks upon unintended human proximity to ensure flawless safety. User-friendly graphical user interface (GUI) or augmented reality (AR) based teaching pendants facilitate rapid reprogramming of robots between projects through demonstration-based learning. Additive manufacturing innovatively 3D prints concrete sand or resin molds evaluating complex architectural flourishes or novel construction methodologies prior to committing resources at scale (Saad et al., 2023). Automated laser scanning and multi-sensor metrology apparatuses precisely verify prefab dimensions and material properties match BIM and quality assurance specifications before logistics. Altogether, proficient robotics and seasoned personnel ingenuity synergistically optimize labor-intensive precasting operations.

Standardized cyber-physical interfaces streamline robots swiftly adapting diverse buildings through dynamic syncing with rich semantic BIM databases. Real-time model alterations permeate plant-wide via factory integrated robotics (FIR) architecture, automatically updating all robot programs, visual aids and processes. Personnel upskill through new technical certificates including robotics, programming and equipment maintenance as manual tasks convert to collaborative automation. Construction firms increasingly outsource precasting to automated plants generating consistent components for multiple concurrent job sites. Visions of the future bring additive construction which extrudes or deposits concrete, composites and metals layer-by-layer into complete structures (Saad et al., 2024). Self-assembling modular robotic subsystems may autonomously configure established buildings from standardized subcomponents.

Economics of Construction Robotics

The global construction sector stands at an inflection point as universal integration of robotics and digitalization revolutionizes productivity. While upfront capital costs of advanced autonomous systems currently exceed human labor rates, return on investment (ROI) analyses indicate rapid cost restoration through optimized operational efficiencies. Continuous manufacturing advances like modular designs, 3D metal printing, and standard component interfaces steadily decrease robot prices through mass scale production and competitive lease-to-own programs (Chen et al., 2022). Innovative firms gaining expertise by testing leading-edge technologies demonstrate compelling payoffs incentivizing followers. Government incentives and public-private research partnerships stimulate initial adoption enterprise wide, recouping investments across expanded tax revenues from economic growth (Xiao et al., 2022). Integrated computer vision and LiDAR scanning conducted by autonomous drones reduce material waste averaging 30% through precision planning, and optimized cutting and assembly processes requiring fewer repeat processes (Mishler and Chen, 2023). A single drone autonomously inspects and maps over 10 times the terrain daily that traditional foot patrols can, pinpointing deviations from plans much earlier in the construction cycle to minimize costly remediation. Powered suits augment workers' strengths, endurance and precision, boosting individual labor productivity from 2 to 3 times per person. Overall, early-adopting firms worldwide report jobsite productivity increases ranging from 20% to 50% through mobile robotics applications (Chen et al., 2022). These gains compound tremendously across large-scale, capital-intensive projects and entire asset portfolio workflows. External support systems and AR-guided tools reduce physical strains and associated injuries, lowering absenteeism over 80% (Wang et al., 2024).

Advance Capital Costs

For any disruptive technology, high-performance systems inevitably face substantial initial expenditures that prevent some conservative early adopters. Within construction, autonomous robots represent pioneering frontier equipment transcending conventional form factors and programmed workflows. Premium pricing currently commissions cutting-edge autonomous capabilities, dexterous manipulators, integrated sensing suites, and modular evolution supporting perpetual refinement (Maxwell et al., 2024). However, costs dramatically compress through normalized design paradigms, mass manufacture economies of scale, and competitive leasing/rental options cultivated by innovation hubs (Rasheed et al., 2023). Standardized mechanical, electrical, and software component interfaces streamline mass customization across diverse manufacturing environments and tasks. Comparable interfaces yield continuity of skills and support structures throughout the technology's lifespan (Maxwell et al., 2024).

Collaborative financing models like competitive lease-to-own programs distributed across portfolios depreciate upfront outlays affordably. Experience demonstrates profitability potential inspiring emulation by progressively bolder organizations (Rasheed et al., 2023). Government programs seeding startup innovation centers promote communities accelerating autonomous construction. Incubators support entrepreneurs commercializing university research through funding, prototyping resources, and business advising. Grant programs offset risks for pilot deployments providing economic value across sectors. Tax incentives stimulate privately financed research partnerships and cooperative skills/infrastructure development between universities, suppliers and builders. Initial public investments directly recoup throughout communities (Siriwardhana and Moehler, 2023).

Robots ultimately achieve cost-competitiveness versus conventional equipment across widening scopes of recurrent, hazardous or unpleasant activities. Mass production economies lower prices into construction mainstream accessibility. While the initial costs are high, they tend to decrease over time as experience with the technology grows, which drives long-term adoption despite initial budget constraints. Contractors strategically sequence purchases maximizing ROI through optimized integration within workflows (Li et al., 2024).

Operational Efficiencies

Robotic deployment significantly enhances operational efficiencies across construction value chains. Mobile autonomous platforms seamlessly cooperating with human experts magnify productivity far beyond pre-automation levels. Computer-integrated machines minimize material wastage through precision workflows optimized via data analytics. Precisely controlled industrial robots reduce cut fiber, sheet metal, lumber and composite scraps averaging 30% or more (Kim et al., 2018). Lasers and waterjets sculpt intricate architectural pieces from designs transmitted digitally. Over time, AI planning eliminates wasteful processes through lean manufacturing principles. On-board analytics flag deviations from planned models, minimizing costly defects, while thermography detects moisture issues before they compromise structural integrity. Personnel swiftly pinpoint faults from aerial surveys rather than ground inspections, allowing project teams to proactively address issues during non-disruptive phases rather than reacting to complaints. Construction robotic systems boost labor output 2 to 3 times by augmenting physical capacities. Motors counterbalance bulky loads, enabling single handed lifting of ordinarily cumbersome materials (Waqar et al., 2024). Sensors prevent muscular overexertion injuries which plague the industry. Computer vision navigates hands-free carrying in cluttered environments. Exo-gloves guide precision fastening with demands exceeding natural dexterities. Reduced fatigue retains experienced personnel amid tight labor markets.

Firms report 20–50% jobsite productivity gains through mobile collaborative robots. Additive manufacturing innovations like large-scale depositing and spraying robots lay foundations, floors and walls with unmatched efficiency and precision. Structural insulated panels fabricated off-site via robotics endure stresses better with air-tight seals versus conventional stick-built methods. Automated just-in-sequence supply logistics fed by real-time demand analytics minimizes on-site material handling (Chen et al., 2022). Drones rerouting deliveries around traffic optimize transportation productivity. Self-driving all-terrain vehicles autonomously navigate worksites mapping in 3D LiDAR for autonomous coordination (Badshah et al., 2024). Remote

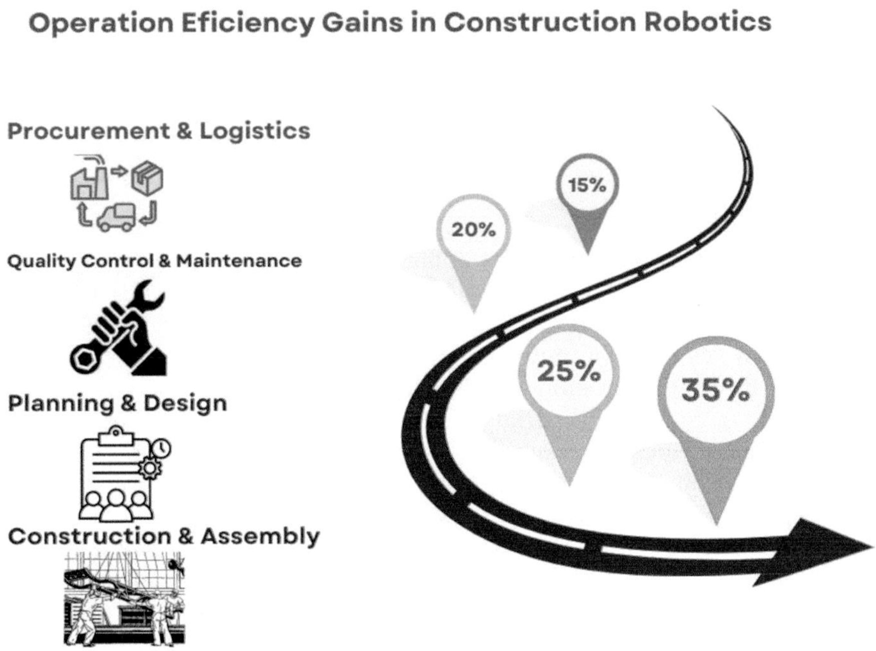

Figure 4: Operational efficiency gains enabled by robotics.

monitoring through mixed reality interfaces links expertise globally for continuous optimization. Together, these technologies empower lifting construction to unprecedented new frontiers. Strategic deployment sequencing maximizes ROI potential by addressing recurring labor-intensive activities. Experienced contractors pilot diverse platforms, cultivating intimate understanding of each technology's capabilities for deriving optimal integration with existing workflows (Zhai and Huang, 2017). Overall, organized human-robot symbiosis elevates entire sector productivity sustainably for generations to come. Figure 4 depicts how efficiency improvements are distributed across different stages of the construction value chain, highlighting the significant role of robotics in planning, procurement, construction, and maintenance. Each portion contributes singularly to the general change, emphasizing the prevalent impact of construction robotics on boosting productivity, decreasing waste, maximizing labor outcomes, and confirming quality and safety.

Risk Mitigation

Beyond immediate operational efficiencies, robotics in construction offer significant risk mitigation benefits that compound economic returns exponentially. Autonomous systems eliminate hazardous tasks like lifts at heights or confined spaces where falls or asphyxiation threaten. Robotic systems and AR guided tools carefully monitor biometrics to prevent muscular strains or repetitive motion injuries. Wearables counterbalance heavy loads or guide precise fastening with millimeter precision far exceeding natural capabilities. Corporates retain skilled labor amid tight markets as health improves (Abdallah et al., 2024). Occupational risks pose previously hidden costs—the average disabling injury costs over $50,000 according to insurers when lost productivity and medical expenses accumulate.

Robotic substitutes for hazardous manual tasks enjoy vast competitive advantages versus human labor costs accounting for direct medical care, insurance premiums, and downtime (Zhou et al., 2022). Merely a handful of prevented severe incidents recoup expensive robot investments. As autonomy increases, risks scale down across portfolios protecting communities and beneficiaries downstream. For example, capabilities mature and upfront costs decrease toward parity with human labor across wider scopes, the returns from risk mitigation justify investments within a shorter timeframe. Continuous progress cycles nurture adoption even with constrained budgets. Strategic long-view investments sow the seeds of optimized, resilient and risk-minimized infrastructure across societies for generations (Zhou et al., 2022). Figure 5 illustrates the varying impacts of construction robotics across different subcategories, emphasizing its considerable role in decreasing exposure to hazard, proactive threat discovery and avoidance, enhanced accident response

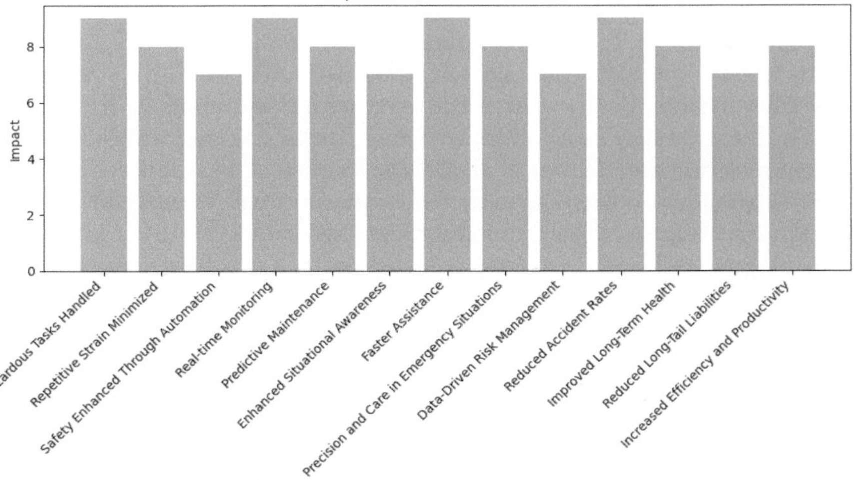

Figure 5: Impact of construction robotics on risk mitigation in construction.

and injury care, and the overall positive effect on lowering accident rates, improving long-term wellness, diminishing long-term financial responsibilities, and boosting efficiency and productivity.

Case Studies: Innovative Robotics in Construction

Leading construction firms driving innovative robotics integration consistently demonstrate groundbreaking productivity gains that inspire broader adoption. Their real-world projects provide compelling performance data validating automation's potential to radically transform industry practices globally. Hadrian Technologies deployed their autonomous bricklaying robot, Hadrian X, on a large mixed-use development in Australia (Yang and Cai, 2024). Outfitted with gripping technology and computer vision for precise positioning, a single Hadrian robot completed exterior walls 25% faster than human crews despite working independently. Interior walls increased twice as quickly versus manual teams (Fu et al., 2024). Hadrian smoothly adapted to architectural variations and material inconsistencies that challenge humans. Ultra-tight mortar control refined from traditional methods reduced material usage by 10–15%, significantly slashing costs. Hadrian's precision eliminated rework, boosting quality. Impressed with initial results, the developer contracted Hadrian for four additional housing communities with deadlines 6–12 months sooner than planned (Liu et al., 2024). Early completions allowed accelerated lease-ups boosting returns versus conventional schedules.

In Tokyo, a major infrastructure firm tasked with the urgent seismic retrofit of a hazardous 500,000 sq ft concrete plant deployed Brokk demolition robots. Operated remotely via virtual reality (VR) interfaces from a mile away, Brokk robots outfitted with shattering teeth and grapples systematically dismantled structural elements from the top down. Integrated sensors identified live utilities to methodically isolate systems beforehand (Zhang et al., 2024). Deconstruction far exceeded goals at 50% greater productivity and zero crew injuries working in enclosed contaminated areas. Project managers credited Brokk with avoiding delays that plague traditional manual contracting in hazardous environments. Municipal authorities keen to accelerate pipeline renovations partnered further testing automated demolition applications.

Meanwhile, in northern Finland, prefabricated Fast Bridge collaborated with Anthropic AI to develop a robotic assembly cell for structural insulated panels (SIPs). Mounted collaborative robots conducted complex SIP cutting, welding and insulating operations with record-low 1.5% waste margins versus typical manual margins of 5–10%. A single cell staffed by 8 robots ran continuously 24/7, tripling Fast Bridge's throughput while adapting to custom variations (Fu et al., 2024). Robotic system-augmented workers installed completed SIPs under computer vision guidance, avoiding cumbersome manual lifting while ensuring highest quality. Buildings leveraging Fast Bridge's mass-custom prefabricated SIPs achieved weather-tight envelope closure 50% faster than stick-built equivalents, supporting increased returns through schedule-dependent value creation. Impressed early adopters rapidly converted 50% of the business to robotic prefabrication based on measurable benefits validated across projects (Yang and Cai, 2024). These innovative case studies quantify construction robotics' potential for unlocking unprecedented economic impacts through higher throughput, optimized resource efficiency, risks mitigation, and reliability improvements. Documented performance provides compelling evidence that energizes the spread of robotics across various related industries.

Autonomous Brick Installation

When Construction Robotics unveiled Hadrian X, the world's first fully autonomous bricklaying robot, skeptics questioned whether robot masonry could match human craftsmanship and productivity. A Denver affordable housing developer sought to validate Hadrian's capabilities on a major project. The 400-unit community's tight construction schedule left little room for delays as housing shortages grew acute (Yang and Cai, 2024). Traditional masonry threatened missing critical deadlines dependent on wall erection. The developer contracted Construction Robotics to deploy Hadrian X across the spreading site. Four Hadrian robots arrived equipped with vacuum-assisted brick handling, computer vision for precise placement, and remote monitoring interfaces for a single operator to oversee multiple robots (Fu et al., 2024). Hadrian mapped architectural plans, seamlessly adaptable to dimensional variability or changes. Trials established optimal brick patterns, cutting overages to a minimum through robotic memory and feedback.

Mason crews laid initial walls for Hadrian to reference. From there, a single Hadrian robot erected exterior walls 25% faster alone versus human teams. Interior walls rose twice as quickly given wider spaces. Hadrian seamlessly integrated around site obstacles from scaffolding to plumbing fixtures (Zhang et al., 2024). Ultra-tight mortar control requiring no troweling wasted 10–15% less material versus manual practices. Per operator supervision, an additional three Hadrian robots joined, maximizing wall production. Together they precisely laid over 1,000 bricks hourly—an unprecedented rate eight times greater than skilled human masons averaging 80–120 bricks per hour. Hadrian worked continuously except brief maintenance periods, achieving up to triple daily work compared to intermittent human crews. Most remarkably, not a single brick required replacement due to Hadrian's hyper-precise placement repeatable to millimeters. Construction Robotics' expertise refined Hadrian eliminating ghosting or other exceptions that trouble initiate masons (Dodampegama et al., 2024). Nearly a year into the job, not one defect surfaced necessitating costly repair, an unheard-of achievement for any method at such a massive scale.

Infrastructure Renovation

The seismic upgrades for an aging concrete plant in Tokyo became critical due to its large reinforced structure, necessitating completion within stringent compliance deadlines. Confined work areas enclosed live utilities amid damaged structures compromised safety. The contractor turned to innovative robotics solutions (Subbotin et al., 2023). The 500,000 square foot facility spans several city blocks, and its reinforced concrete construction compounds renovation difficulties. Traditional dismantlement posed risks of mishaps incurring fines or injuries delaying the restoration of critical infrastructure. The contractor partnered with Komatsu to deploy excavators retrofitted with advanced robotic demolition arms.

Komatsu integrated force-detecting sensors, 3D scanning, and millimeter-precision manipulation into hydraulic excavator booms (Hee Chung et al., 2023). Remote operation interfaces covered in virtual reality headsets project demolition in workspace scale from a safe distance. Simulations optimized dismantling sequences minimizing risks while sustaining the surroundings. Robotic demolition commenced with explosives experts containing charges within structural members, pre-weakening them for excavators. Remotely activated hydraulic shears effortlessly cut through steel in occupied zones impervious to fatigue (Sirimewan et al., 2024). Scanners identified live utilities, segregating circuits before delicately dismantling around them. Force sensors detected unexpected obstructions like embedded reinforcements, navigating contactless diversion. Scans crosschecked deconstruction against as-built renderings in real-time, adapting to irregularities. Material deposition sorted rubble streams for efficient recycling versus dust clouds endangering air quality compliance (Dominguez et al., 2024). Unmanned round-the-clock operation tripled initial productivity estimates versus restricted manual shifts. Excavators labored continuously except for brief maintenance, demolishing mountainous areas completed in months by traditional crews within only weeks. Zero safety incidents occurred from standoff automation eliminating exposure to hazardous solvents or cave-ins.

Moreover, schedule compression liberated critical resources. Neighboring lots repurposed storage offset logistics bottlenecks, unblocking traffic flows (Hee Chung et al., 2023). Accelerated rebar fabrication and concrete pouring commenced in cleared zones several months ahead of schedule. Expedited civil works enabled structural work starting concurrently versus sequentially dependent origination. Radiant, the municipal authority commissioned an additional phase leveraging Komatsu robots for infrastructure-dense pipeline replacements. Underground demolition compressed from years to only fifteen months slashing disruptions. Cost-effective automation establishes recurring services converting Japan's aging built environment sustainably (Subbotin et al., 2023). Restored on schedule, the industrial plant resumed operations, giving a boost to the local economy.

Reinforced Concrete Construction

The redevelopment of Battersea Power Station, a beloved London landmark, faced immense challenges to convert the hulking reinforced concrete structures into a new mixed-use destination safely and economically (Liu et al., 2024). A 20,000-ton demolition entitled particular caution to sustain surrounding architecture. Project managers evaluated traditional dismantlement risking multi-year delays. Confined spaces exposed manual crews to asbestos and collapse dangers. A safer robotic alternative emerged through Brokk's innovative remote-controlled demolition robots.

Brokk robots equipped with hydraulic breakers were deployed to demolish resilient concrete. They utilized 360-degree cameras to navigate tight spaces. Dual-arm manipulators meticulously excavated structural members given millimeter precision. Force-feedback control contact ensured the protection of heritage architecture during the process (Leder and Menges, 2023). Collaborating with explosive experts, Brokk robots precisely targeted pre-weakened zones. Charges embedded within structural sections pre-fractured debris, allowing hydraulic shears to operate without causing collateral damage. The robots coordinated movements sequencing their movements to sequence complete removals and prevent overflow. Project managers remotely controlled multiple Brokk robots simultaneously using VR interfaces (Shayesteh et al., 2023). Panoramic views displayed the demolition progress from a safe distance of one mile. Real-time scanning modeled the advancing deconstruction, adjusting cleaning sequences to preserve intact facades. Round-the-clock Brokk shifts far outpaced manual crews. A single robot matched 200 workers' monthly output within only weeks. Unmanned robots labored continuously except brief servicing, altogether eliminating safety delays and achieving a rate over 4 times faster than projected for human crews. Robotic demolition significantly shortened the original 16-month schedule by several crucial months. Accelerated clearing liberated earlier foundation work, enabling concurrent versus sequential dependency (Oke et al., 2023). Advanced structural erection started months ahead of schedule dependent on conventional clearance. Notably, confined twisting chambers posed severe hazards manually yet yielded most valuable materials for reuse or recycling (Halder et al., 2023).

Tunnel Infrastructure

Toronto's York Spadina Subway Extension faced barriers in completing tunnels on schedule and budget. Each tunnel required 16,000 custom concrete lining segments, each weighing up to 5 tons, which were traditionally installed manually in confined spaces. Anthropic partnered with the tunneling contractor to develop an automated production cell leveraging collaborative robots (Damián and Zamorano, 2022).

In a factory setting, robotic cutters, welders, and bolters were deployed to prefabricate segments at an unmatched rate. These robotic systems complemented workers by precisely handling segments within the factory. Counterbalanced to mitigate fatigue, workers naturally lifted and positioned segments for robotic fabrication 2 to 3 times an individual's capacity (Dai et al., 2023). Computer vision and force feedback guided operations ergonomically within the autonomous cell. Round-the-clock operation doubled previous segment outputs with fewer workers. Robots adapted cut angles and assembly sequences to variable segment designs for rail curves without impacting precision or quality. Waste margins reduced from typical 5–10% tolerances during precasting to only 1.5% optimized fabrication.

Meanwhile, tunnel reinforcements installed excavation support immediately following tunnel boring machines to accelerate schedules (Li et al., 2024). Early reinforcements stabilized freshly exposed soil, permitting simultaneous excavation and lining installation versus sequential dependent methods slowing legacy projects. Robotic installation cells transported finished segments autonomously to tunnel fronts. Robotic-assisted installers accurately aligned interlocking joints for automated welders, bolters and sealers to permanently secure with precision far surpassing human variability. Robot fault tolerance and calibration ensured consistent hermetic installations (Bellini Machado and Massao Futai, 2024). Round-the-clock robotic installation outpaced tunneling by 25% versus conventional lags improving safety. Confined tunnels no longer demanded disruptive weekend shutdowns for manual lining erection, keeping excavations continuously waterproof. Automation eliminated the need for scaffolding and access challenges that hamper productivity (Dai et al., 2023). Accurate joints eliminated pressure leaks and settlement risks compromising tunnel integrity over decades. Savings from optimized installations and accelerated schedules significantly amplified the economic impact on the construction sector. Robotically upgraded tunnels delivered on time under expanding budgets versus costly delays. Anthropic's success validated autonomous production and installation for tunnel infrastructure globally. Robotic innovations eliminated confined space hazards while multiplying outputs, optimizing designs, and accelerating schedules sustainably for generations to benefit (Li et al., 2024). Innovative implementations have paved the way for automation, revolutionizing heavy civil construction globally.

References

Abdallah, A. M., Badawi, M., Elsamak, G., Hu, J. W., Mlybari, E. A. et al. (2024). Strengthening of RC beams with inadequate lap splice length using cast-*in-situ* and anchored precast ECC ferrocement layers mitigating construction failure risk. *Case Studies in Construction Materials,* 20: e02747. https://doi.org/10.1016/j.cscm.2023.e02747.

Ali, Z., Saad, S., Rasheed, K. and Ammad, S. (2024). AI future perspectives and trends in construction. In *AI in Material Science* (pp. 239–261). CRC Press. https://doi.org/10.1201/9781003438489-11.

Ammad, S., Saad, S., Bashir, M. T., Qureshi, A. H., Altaf, M. et al. (2021). Construction accidents via integrating Building Information Modelling (BIM) with emerging digital technologies a review. *2021 3rd International Sustainability and Resilience Conference: Climate Change*. https://doi.org/10.1109/IEEECONF53624.2021.9668066.

Anane, W., Iordanova, I. and Ouellet-Plamondon, C. (2022). Modular robotic prefabrication of discrete aggregations driven by BIM and computational design. *Procedia Computer Science,* 200: 1103–1112. https://doi.org/10.1016/j.procs.2022.01.310.

Badshah, I., Saad, S., Ammad, S., Rasheed, K., Khan, M. W. et al. (2024). Enhancing sustainable mobility: Lane change strategies for efficient traffic flow at stationary bottlenecks. *2024 ASU International Conference in Emerging Technologies for Sustainability and Intelligent Systems (ICETSIS),* 1197–1202. https://doi.org/10.1109/ICETSIS61505.2024.10459583.

Baduge, S. K., Thilakarathna, S., Perera, J. S., Arashpour, M., Sharafi, P. et al. (2022). Artificial intelligence and smart vision for building and construction 4.0: Machine and deep learning methods and applications. *Automation in Construction,* 141: 104440. https://doi.org/https://doi.org/10.1016/j.autcon.2022.104440.

Bellini Machado, L. and Massao Futai, M. (2024). Tunnel performance prediction through degradation inspection and digital twin construction. *Tunnelling and Underground Space Technology,* 144: 105544. https://doi.org/10.1016/j.tust.2023.105544.

Chen, C., Li, X., Yao, W., Wang, Z., Zhu, H. et al. (2022). Analysis of the impact of construction robots on workers' health. *Building and Environment,* 225: 109595. https://doi.org/10.1016/j.buildenv.2022.109595.

Dai, X., Bidarmaghz, A. and Narsilio, G. A. (2023). Energy tunnels: A review of the state of the art and knowledge gaps to harness renewable energy from underground infrastructure. *Tunnelling and Underground Space Technology,* 142: 105431. https://doi.org/10.1016/j.tust.2023.105431.

Damián, R. and Zamorano, C. I. (2022). Environmental impact assessment of high-speed railway tunnel construction: A case study for five different rock mass rating classes. *Transportation Geotechnics*, 36: 100817. https://doi.org/10.1016/j.trgeo.2022.100817.

Dodampegama, S., Hou, L., Asadi, E., Zhang, G., Setunge, S. et al. (2024). Revolutionizing construction and demolition waste sorting: Insights from artificial intelligence and robotic applications. *Resources, Conservation and Recycling*, 202: 107375. https://doi.org/10.1016/j.resconrec.2023.107375.

Dominguez, C., Kakkos, E., Gross, D., Hischier, R., Orehounig, K. et al. (2024). Renovated or replaced? Finding the optimal solution for an existing building considering cumulative CO2 emissions, energy consumption and costs—A case study. *Energy and Buildings*, 303: 113767. https://doi.org/10.1016/j.enbuild.2023.113767.

Dui, H., Xu, H., Zhang, L. and Wang, J. (2023). Cost-based preventive maintenance of industrial robot system. *Reliability Engineering & System Safety*, 240: 109595. https://doi.org/10.1016/j.ress.2023.109595.

Filippi, E., Bannò, M. and Trento, S. (2023). Automation technologies and their impact on employment: A review, synthesis and future research agenda. *Technological Forecasting and Social Change*, 191: 122448. https://doi.org/10.1016/j.techfore.2023.122448.

Fu, Y., Chen, J. and Lu, W. (2024). Human-robot collaboration for modular construction manufacturing: Review of academic research. *Automation in Construction*, 158: 105196. https://doi.org/10.1016/j.autcon.2023.105196.

Halder, S., Afsari, K., Chiou, E., Patrick, R., Hamed, K. A. et al. (2023). Construction inspection & monitoring with quadruped robots in future human-robot teaming: A preliminary study. *Journal of Building Engineering*, 65: 105814. https://doi.org/10.1016/j.jobe.2022.105814.

Haris, M., Saad, S., Ammad, S. and Rasheed, K. (2024). AI in fabrication and construction. In *AI in Material Science* (pp. 169–192). CRC Press. https://doi.org/10.1201/9781003438489-8.

Hee Chung, M., Eun Kim, S., Woo Song, Y. and Chul Park, J. (2023). Strategies for improving impact of energy renovation: A case study on Korean daycare centers. *Energy and Buildings*, 284: 112844. https://doi.org/10.1016/j.enbuild.2023.112844.

Huang, L., Cai, W., Zhu, Z. and Zou, Z. (2023). Dexterous manipulation of construction tools using anthropomorphic robotic hand. *Automation in Construction*, 156: 105133. https://doi.org/10.1016/j.autcon.2023.105133.

Iturralde, K., Pan, W., Linner, T. and Bock, T. (2022). Automation and robotic technologies in the construction context: Research experiences in prefabricated façade modules. In *Rethinking Building Skins* (pp. 475–493). Elsevier. https://doi.org/10.1016/B978-0-12-822477-9.00009-7.

Jiang, P., Wang, Z., Li, X., Wang, X. V., Yang, B. et al. (2023). Energy consumption prediction and optimization of industrial robots based on LSTM. *Journal of Manufacturing Systems*, 70: 137–148. https://doi.org/10.1016/j.jmsy.2023.07.009.

Kaur, B., Mansi Dimri, S., Singh, J., Mishra, S., Chauhan, N. et al. (2023). Insights into the harvesting tools and equipment's for horticultural crops: From then to now. *Journal of Agriculture and Food Research*, 14: 100814. https://doi.org/10.1016/j.jafr.2023.100814.

Khan, M. I., Khan, N., Saad, S. and Khan, M. W. (2024). Artificial Intelligence in construction materials. In *AI in Material Science* (pp. 47–64). CRC Press. https://doi.org/10.1201/9781003438489-3.

Kim, T., Lee, U. -K., Kim, S. W., Lim, H., Kim, C. -W. et al. (2018). Flexible double-cage hoist for high operational efficiency in tall building construction. *Automation in Construction*, 96: 280–291. https://doi.org/10.1016/j.autcon.2018.09.023.

Leder, S. and Menges, A. (2023). Architectural design in collective robotic construction. *Automation in Construction*, 156: 105082. https://doi.org/10.1016/j.autcon.2023.105082.

Li, J., Wu, Z., Yu, K. and Zhao, W. (2024). The effect of industrial robot adoption on firm value: Evidence from China. *Finance Research Letters*, 60: 104907. https://doi.org/10.1016/j.frl.2023.104907.

Li, T., Li, X., Rui, Y., Ling, J., Zhao, S. et al. (2024). Digital twin for intelligent tunnel construction. *Automation in Construction*, 158: 105210. https://doi.org/10.1016/j.autcon.2023.105210.

Liu, Y., Alias, A. H. bin, Haron, N. A., Bakar, N. A., Wang, H. et al. (2024). Technology status tracing and trends in construction robotics: A patent analysis. *World Patent Information*, 76: 102259. https://doi.org/10.1016/j.wpi.2023.102259.

Maxwell, B. M., Christianson, R. D., Arch, R., Johnson, S., Book, R. et al. (2024). Applied denitrifying bioreactor cost efficiencies based on empirical construction costs and nitrate removal. *Journal of Environmental Management*, 352: 120054. https://doi.org/10.1016/j.jenvman.2024.120054.

Mishler, S. and Chen, J. (2023). Effect of automation failure type on trust development in driving automation systems. *Applied Ergonomics*, 106: 103913. https://doi.org/10.1016/j.apergo.2022.103913.

Naeem, U. H., Bilal, M., Syed, F., Rasheed, K., Saad, S. et al. (2022). A multilayer encryption model to protect healthcare data in cloud environment. *2022 International Conference on Data Analytics for Business and Industry, ICDABI 2022*. https://doi.org/10.1109/ICDABI56818.2022.10041708.

Oke, A. E., Aliu, J., Oluwasefunmi Fadamiro, P., Akanni, P. O., Stephen, S. S. et al. (2023). Attaining digital transformation in construction: An appraisal of the awareness and usage of automation techniques. *Journal of Building Engineering*, 67: 105968. https://doi.org/10.1016/j.jobe.2023.105968.

Rasheed, K., Saad, S. and Ammad, S. (2023). Green financing for sustainable economic growth: Exploring critical factors towards resilient economies. *2023 International Conference on Sustainable Islamic Business and Finance (SIBF)*, 104–111. https://doi.org/10.1109/SIBF60067.2023.10379984.

Rasheed, K., Saad, S., Ammad, S. and Bashir, M. T. (2024). Industry 4.0 and construction. In *AI in Material Science* (pp. 65–87). CRC Press. https://doi.org/10.1201/9781003438489-4.

Rasheed, K., Saad, S., Ammad, S., Iftikhar Ali, M., Bilal, M. et al. (2023). Towards sustainable growth management: Analyzing the consequences of ecological deterioration in asian countries on energy consumption, economic growth and financial progress. *2023 International Conference on Sustainable Islamic Business and Finance (SIBF)*, 31–36. https://doi.org/10.1109/SIBF60067.2023.10379892.

Rasheed, K., Saad, S., Shahzad, L., Ammad, S., Ali, A. et al. (2021). Application of gesture data recognition in a human-interactive leap motion sensor chair. *2021 International Conference on Data Analytics for Business and Industry, ICDABI 2021*. https://doi.org/10.1109/ICDABI53623.2021.9655819.

Rasheed, K., Waqas, M., Saad, S., Zaland, A., Ali, Z. et al. (2024). Ethical and societal implications of AI in construction. In *AI in Material Science* (pp. 216–238). CRC Press. https://doi.org/10.1201/9781003438489-10.

Rasheed, K., Zaland, A., Saad, S., Ammad, S., Rostami, A. et al. (2024). History of AI. In *AI in Material Science* (pp. 15–46). CRC Press. https://doi.org/10.1201/9781003438489-2.

Saad, S., Ammad, S. and Rasheed, K. (2024). *AI in Material Science*. CRC Press. https://doi.org/10.1201/9781003438489.

Saad, S., Ammad, S., Rasheed, K., Balbehaith, M., Qureshi, A. H. et al. (2023). Enhancing construction management in restoration through BIM-assisted photogrammetric 3D Model Accuracy Assessment. *2023 International Conference on Sustaining Heritage: Innovative and Digital Approaches (ICSH)*, 100–103. https://doi.org/10.1109/ICSH57060.2023.10482845.

Saad, S., Haris, M., Ammad, S. and Rasheed, K. (2024). AI-assisted building design. In *AI in Material Science* (pp. 143–168). CRC Press. https://doi.org/10.1201/9781003438489-7.

Saad, S., Rasheed, K., Ammad, S., Ali, Z., Zaland, A. et al. (2024). The conclusive role of AI in Industry 4.0 construction materials. In *AI in Material Science* (pp. 262–270). CRC Press. https://doi.org/10.1201/9781003438489-12.

Saad, S., Rasheed, K., Ammad, S., Waqas Khan, M., Zaland, A. et al. (2024). Is AI the architect of tomorrow's materials in the age of Industry 4.0? In *AI in Material Science* (pp. 1–14). CRC Press. https://doi.org/10.1201/9781003438489-1.

Saad, S., Zaland, A., Ammad, S. and Rasheed, K. (2024). Robotics in construction. In *AI in Material Science* (pp. 115–142). CRC Press. https://doi.org/10.1201/9781003438489-6.

Shayesteh, S., Ojha, A., Liu, Y. and Jebelli, H. (2023). Human-robot teaming in construction: Evaluative safety training through the integration of immersive technologies and wearable physiological sensing. *Safety Science*, 159: 106019. https://doi.org/10.1016/j.ssci.2022.106019.

Sirimewan, D., Bazli, M., Raman, S., Mohandes, S. R., Kineber, A. F. et al. (2024). Deep learning-based models for environmental management: Recognizing construction, renovation, and demolition waste in-the-wild. *Journal of Environmental Management*, 351: 119908. https://doi.org/10.1016/j.jenvman.2023.119908.

Siriwardhana, S. and Moehler, R. C. (2023). Enabling productivity goals through construction 4.0 skills: Theories, debates, definitions. *Journal of Cleaner Production*, 425: 139011. https://doi.org/10.1016/j.jclepro.2023.139011.

Subbotin, A. S., Konchakov, D. I., Karavashkina, E. A. and Dolzhko, D. B. (2023). Development of electric transport and construction of transport infrastructure in the conditions of design for the renovation of small towns. *Transportation Research Procedia*, 68: 654–658. https://doi.org/10.1016/j.trpro.2023.02.090.

Sun, Y., Jeelani, I. and Gheisari, M. (2023). Safe human-robot collaboration in construction: A conceptual perspective. *Journal of Safety Research*, 86: 39–51. https://doi.org/10.1016/j.jsr.2023.06.006.

Tapia, J., Knoop, E., Mutný, M., Otaduy, M. A., Bächer, M. et al. (2020). MakeSense: automated sensor design for proprioceptive soft robots. *Soft Robotics*, 7(3): 332–345. https://doi.org/10.1089/soro.2018.0162.

van der Heijden, J. (2023). Construction 4.0 in a narrow and broad sense: A systematic and comprehensive literature review. *Building and Environment*, 244: 110788. https://doi.org/10.1016/j.buildenv.2023.110788.

Wang, W., Zaheer, Q., Qiu, S., Wang, W., Ai, C., Wang, J., Wang, S. and Hu, W. (2024). Digital Twins technologies. In *Digital Twin Technologies in Transportation Infrastructure Management* (pp. 27–74). Springer Nature Singapore. https://doi.org/10.1007/978-981-99-5804-7_2.

Waqar, A., Mateen Khan, A. and Othman, I. (2024). Blockchain empowerment in construction supply chains: Enhancing efficiency and sustainability for an infrastructure development. *Journal of Infrastructure Intelligence and Resilience*, 3(1): 100065. https://doi.org/10.1016/j.iintel.2023.100065.

Waqas Khan, M., Saad, S., Ammad, S., Rasheed, K., Jamal, Q. et al. (2024). Smart infrastructure and AI. In *AI in Material Science* (pp. 193–215). CRC Press. https://doi.org/10.1201/9781003438489-9.

Xiao, B., Chen, C. and Yin, X. (2022). Recent advancements of robotics in construction. *Automation in Construction*, 144: 104591. https://doi.org/https://doi.org/10.1016/j.autcon.2022.104591.

Yang, L. and Cai, H. (2024). Enhanced visual SLAM for construction robots by efficient integration of dynamic object segmentation and scene semantics. *Advanced Engineering Informatics*, 59: 102313. https://doi.org/10.1016/j.aei.2023.102313.

Yevu, S. K., Owusu, E. K., Chan, A. P. C., Sepasgozar, S. M. E., Kamat, V. R. et al. (2023). Digital twin-enabled prefabrication supply chain for smart construction and carbon emissions evaluation in building projects. *Journal of Building Engineering*, 78: 107598. https://doi.org/10.1016/j.jobe.2023.107598.

Zaland, A., Saad, S., Ahmad, W., Ammad, S., Khan, M. W. et al. (2024). Enhancing mechanical properties of concrete by adding polypropylene fibers. *2024 ASU International Conference in Emerging Technologies for Sustainability and Intelligent Systems (ICETSIS)*, 771–778. https://doi.org/10.1109/ICETSIS61505.2024.10459663.

Zaland, A., Saad, S., Rasheed, K. and Ammad, S. (2024). Automation in construction industry. In *AI in Material Science* (pp. 88–114). CRC Press. https://doi.org/10.1201/9781003438489-5.

Zhai, Y. and Huang, G. Q. (2017). Operational hedging and coordination in prefabrication construction industry. *Procedia Manufacturing*, 11: 1178–1183. https://doi.org/10.1016/j.promfg.2017.07.242.

Zhang, M., Xu, R., Wu, H., Pan, J., Luo, X. et al. (2023). Human–robot collaboration for on-site construction. *Automation in Construction*, 150: 104812. https://doi.org/10.1016/j.autcon.2023.104812.

Zhang, X., Wang, Y., Li, C., Fahmy, A., Sienz, J. et al. (2024). Innovative multi-dimensional learning algorithm and experiment design for human-robot cooperation. *Applied Mathematical Modelling*. https://doi.org/10.1016/j.apm.2024.01.001.

Zhao, X., Jin, Y., Selvaraj, N. M., Ilyas, M., Cheah, C. C. et al. (2023). Platform-independent visual installation progress monitoring for construction automation. *Automation in Construction*, 154: 104996. https://doi.org/10.1016/j.autcon.2023.104996.

Zhou, Z., Goh, Y. M., Shi, Q., Qi, H., Liu, S. et al. (2022). Data-driven determination of collapse accident patterns for the mitigation of safety risks at metro construction sites. *Tunnelling and Underground Space Technology*, 127: 104616. https://doi.org/10.1016/j.tust.2022.104616.

Chapter 5

Digital Twin and Sustainability

Nexus for Constructive Impact

Ahmad Zaland,[1] *Habib Ullah,*[2] *Kumeel Rasheed,*[1,*]
Syed Saad[3] and *Syed Ammad*[4]

Introduction

In the rapidly evolving technological landscape, digital twins have ushered in a new era of creativity and efficiency across various industries. These digital copies of real-world items, operations, or systems have moved beyond theory and are now significantly influencing how we live our daily lives. Manufacturing is one of the sectors where digital twins have the most tangible effects. In this industry, virtual product models are made possible by technology, which helps businesses optimize their production and design processes. Time-to-market and production costs are significantly reduced as a result. With the use of digital twins, producers can ensure a more efficient and less expensive manufacturing process by simulating, analyzing, and troubleshooting possible problems in a virtual environment before physical production starts (Soori et al., 2023). Digital twins are transforming patient care in healthcare by enabling providers anticipate future health problems, tailor treatment strategies, and investigate the efficacy of various approaches through virtual versions of individual patients. In addition to improving patient outcomes, this tailored strategy makes healthcare delivery more effective and efficient (Katsoulakis et al., 2024).

[1] Department of Civil Engineering, CECOS University of IT and Emerging Sciences, Pakistan.
[2] School of Architecture, Southeast University, Sipailou Campus, China.
[3] Science, Engineering & Manufacturing Centre Sandwell College West Bromwich, United Kingdom.
[4] School of Engineering and Built Environment, Griffith University, Gold Coast, Australia.
* Corresponding author: kumeelrasheed@cecosian.edu.pk

Integrating digital twins into smart cities has profound implications. These digital replicas of urban environments enable city planners to model and optimize various scenarios from energy consumption to traffic management. The result is a more resilient, responsive, and sustainable urban infrastructure that can adapt to meet the evolving needs of its inhabitants (Hu, 2023). Digital twins significantly impact various industries in real life. They have demonstrated their transformative potential, driving effectiveness, innovation, and a more integrated future across diverse fields such as healthcare, manufacturing and urban development. As technology advances digital twins will become increasingly prevalent bringing leading to further significant and beneficial enhancements in our daily lives (Attaran and Celik, 2023).

Understanding the Relationship

Digital twins are advanced virtual models of physical systems, realities, or processes that exist within a computer environment. These models have garnered significant attention due to their ability to offer real-time monitoring, enhanced functional insights, and predictive capabilities. By linking the digital and physical worlds, digital twins provide a dynamic and interactive representation of each other. This revolutionary technology is being utilized widely to boost productivity, reduce costs, and spur innovation.

Definition and Components of Digital Twins

A digital twin consists of several key components that collaborate to create an accurate and functional virtual replica. This includes the real-world physical system or object as illustrated in Figure 1. The digital replica mirrors its physical counterpart, ensuring precision and utility.

- **Virtual Model:** A 3D model, simulation, or data visualization that is a computerized depiction of the real world (Zhang, 2018).
- **Data:** Precise and instantaneous information gathered from the tangible world through sensors, Internet of Things bias, and additional data sources (Elgazzar et al., 2022).
- **Connectivity:** The framework for communication that permits data to continuously flow across the digital and physical worlds (Wook and Mo, 2024).
- **Analytics:** The instruments and methods for repurposing and analyzing the data, providing insight and foresight (Annaram, 2022).
- **User Interface:** The interface through which users interact with the digital twin, allowing them to monitor, control, and make informed decisions.

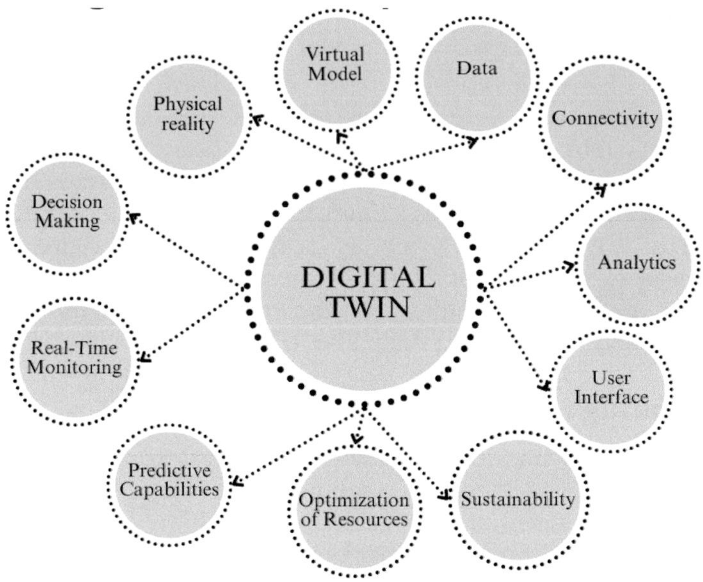

Figure 1: Components and roles of Digital Twins.

Role of Digital Twins in Decision-Making

The primary advantage of digital twins is their potential to facilitate better informed decision-making processes. A digital twin offers a current and thorough view of the status and performance of assets by continuously gathering and analyzing data from IoT detectors and other data sources. Through this real-time feedback loop, associations can predict implicit failures, pinpoint inefficiencies, and enhance functional performance. Simulating various scenarios and predicting issues before implementing real-world changes significantly reduces risks and improves strategic planning (Attaran and Celik, 2023). The constant stream of data makes it possible to identify anomalies, performance problems, and implicit failures immediately. For example, in the manufacturing industry, digital twins can monitor equipment and operations, providing alerts when maintenance is needed or when performance deviates from expected parameters. This innovative method ensures smooth operations, reduces downtime, and extends the lifespan of systems (Bofill et al., 2023). By reducing energy consumption and waste, this approach not only boosts productivity but also supports environmental goals. A factory's digital twin, for instance, can simulate several production schedules and workflows to determine which ones are the most effective in terms of waste reduction and energy consumption (Leng et al., 2021).

Predictive Capabilities

Predictiveness is one of the most potent attributes of digital twins. Digital twins can predict future behavior and results by evaluating performance indicators and past data. This ability to forecast the future is extremely useful for minimizing operating risks, increasing production efficiency, and optimizing maintenance schedules. Digital twins, for example, can be used in the energy sector to ensure continuous and effective energy production by anticipating equipment breakdowns and optimizing maintenance schedules (Zhong et al., 2023).

Optimization of Resources

A digital twin has a significant role in resource optimisation, one of the most crucial aspects of sustainability. It provides a thorough perception into the functioning of physical systems, which facilitates the discovery of inefficiencies and resource depletion. This capability is particularly helpful in industries like manufacturing, construction, and municipal structure operation where resource optimization can significantly improve the terrain and the level of frugal living (He and Mao, 2023). For example, a digital twin of a house can analyze patterns of energy consumption and recommend changes to HVAC (heating, ventilation, and air exertion) systems to improve energy efficiency (Zhang et al., 2024).

Sustainability: A Critical Concern

It is crucial to comprehend the connection between a digital twin and sustainability as it becomes a less pressing issue for governments and corporations globally. The concept of sustainability pertains to the ability to fulfill current needs without compromising the capacity of future generations to fulfill theirs. Its vibrant boundaries include social justice, economic feasibility, and environmental preservation. Sustainability in artificial operations and municipal development settings emphasizes minimizing environmental consequences, making the best use of available resources, and encouraging long-term adaptability (Nativi et al., 2021). Digital twins assist organizations in determining areas where they can lessen their environmental effect by offering real-time data on resource usage, emissions, and energy consumption. An industrial plant's digital twin, for instance, can track energy consumption in real time and recommend changes to lower emissions and consumption (Yu et al., 2022).

Economic Viability

One essential component of sustainability is profitable viability. Digital twins optimize functional effectiveness and cut expenses to assist associations in achieving profitable sustainability. Digital twins considerably lower functional costs by improving product processes, forecasting conservation requirements, and optimizing resource use. For example, a transportation network's digital twin can optimize schedules and routes, saving money on conservation and energy use (Franciosi et al., 2022). Research has indicated that digital twins have the potential to drastically lower carbon emissions and improve operational efficiency. (Tao et al., 2019) emphasized that by enabling more accurate management and optimization of artificial processes, digital twins could effect significant reductions in energy consumption and emigrations. Moreover, (Negri et al., 2017) demonstrated how digital twins could improve resource management and functional efficacy in vibrant industries, supporting sustainability claims. Similarly, the digital twin can aid in anticipating and assuaging implicit environmental risks. Digital twins assist in the development of more adaptable and sustainable designs by testing implicit effects and various scripts.

Digital Twins Driving Sustainable Practices

The incorporation of digital twins into diverse businesses signifies a noteworthy progression in the endeavor to adopt sustainable practices. Digital twins are virtual versions of actual systems that enable thorough asset and process optimization. Figure 2 illustrates that this technology offers practical insights that can result in better resource management, less of an adverse effect on the environment, and increased sustainability in a variety of industries.

Digital Twins in Manufacturing

Digital Twin Applications in Manufacturing

Digital twins have the capability to simulate and optimize whole production lines in the manufacturing industry. Businesses are able to track, evaluate, and enhance their operations in real time by establishing a digital replica of the physical industrial environment. This expertise is essential for locating inefficiencies, cutting waste, and maximizing energy use all of which lead to increased sustainability and production (Bécue et al., 2020).

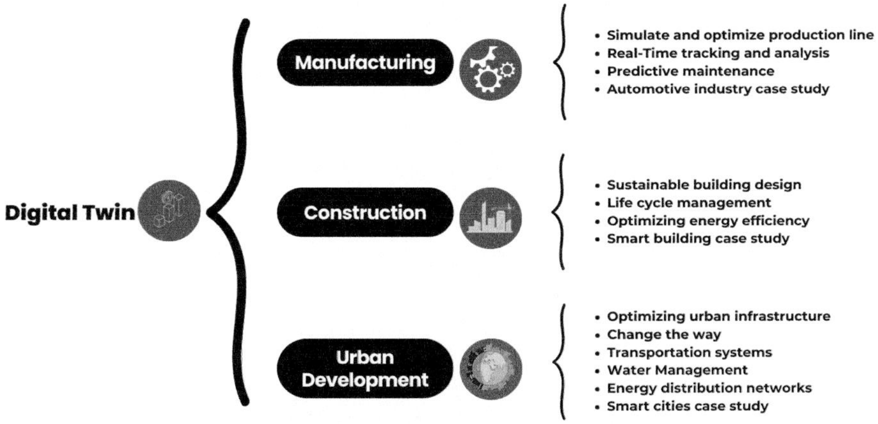

Figure 2: Digital Twin sustainable practices.

Optimizing Production Processes

With the use of digital twins, manufacturers can model various production scenarios and evaluate how they affect productivity and resource usage. For instance, they are able to determine the most energy-efficient configuration by analyzing the energy usage of different manufacturing line setups. This minimizes operating costs in addition to the overall energy footprint. According to (Schleich et al., 2017), digital twins make it possible to precisely regulate and optimize production processes, which significantly reduces waste and energy consumption.

Predictive Maintenance and Reduced Downtime

Prophecy conservation is another essential function of digital twins in production. Digital twins can predict when maintenance is required by continuously monitoring the condition of machinery and equipment, preventing unexpected breakdowns and reducing downtime (Saad et al., 2022). This innovative approach to maintenance not only extends the lifespan of resources but also ensures that production processes operate smoothly, thereby strengthening sustainability efforts. In promoting digital binary technology, the automobile industry has been colonial. Digital twins are used by big automakers to test, create, and optimize their product processes. For example, digital twins can simulate a full assembly line, enabling geniuses to experiment with many setups and determine the most efficient and long-lasting product designs. Automakers can minimize waste, cut energy use, and lessen their carbon footprint by streamlining these procedures (Papanagnou, 2020).

Digital Twins in Construction

Sustainable Building Design and Construction

Another industry where digital twins significantly impact sustainability is construction. Engineers and architects can optimize a structure's design, building, and functional phases with the help of digital twins. Through strategic manipulation of multiple design blueprints, digital twins allow for the selection of the most sustainable features and building methods, minimizing the ecological footprint of new constructions (Saad et al., 2024).

Lifecycle Management of Buildings

Digital twins play a crucial role throughout the lifecycle of a structure existence, not just during the design and construction stages. By embedding IoT sensors digital twins enable continuous monitoring of a structure's performance, including water usage, energy consumption, and interior environmental quality (Yang et al., 2022). The continuous optimization of structural operations, enabled by real-time data, ensures the efficiency and sustainability.

Optimizing Energy Efficiency

One of the primary benefits of digital twins in construction is optimizing energy efficiency. Structural managers can identify optimal strategies to reduce energy consumption through various energy management techniques. Digital twin, for instance, can simulate the effects of various HVAC configurations and recommend adjustments to enhance efficiency. This minimizes the structure's functioning costs as well as its carbon footprint.

Smart Buildings

An exemplary instance of how digital elements can enhance sustainability in the construction sector is smart structures. These systems encompass and oversee diverse components like heating, cooling, and lighting using digital twins. Digital twins have the ability to assess real-time data and optimize these systems' performances, ensuring that energy is used sustainably and efficiently (Waqas Khan et al., 2024). For example, digital twins can adjust lighting based on occupancy, reducing energy waste and optimizing overall energy efficiency.

Digital Twins in Urban Development

Optimizing Urban Infrastructure

Digital twins are also very beneficial for civic development. Digital twins can be used by city planners and authorities to model and optimize the civic structure, which includes networks for energy distribution, water operations, and transportation. Digital twins assist in forming data-driven decisions that support sustainability and adaptation by bluffing the effects of various development scripts.

Transportation Systems

Transport systems can be modeled and optimized by digital twins to increase efficiency and lessen their impact on the environment (Jafari et al., 2023). They can, for instance, make up business patterns and determine the best strategies for lowering traffic and emigration. Digital twins can assist metropolises in lowering their carbon footprint and improving the general standard of living for residents by streamlining corporate traffic and public transport timetables (Badshah et al., 2024).

Water Management

Water operations is another crucial area where digital twins might promote sustainability is in. They can further effectively operate water offers by identifying leaks and inefficiencies by modeling water distribution networks. This is especially crucial in areas experiencing water scarcity, as sustainable development depends on efficient water management.

Energy Distribution Networks

Digital twins are also crucial for optimizing energy distribution networks. They can assist metropolises in better balancing supply and demand, reducing energy waste, and integrate renewable energy sources by optimizing various energy operating systems (Rasheed et al., 2023). For example, can forecast the impact of wind and solar energy on electrical systems and suggest optimal locations to install them.

Smart Cities

Smart cities are leading the way in leveraging digital twins to promote sustainability. These cities have an impact on digital twin management and optimization of vibrant municipal systems, ranging from waste management and public services to energy and transportation (Haris et al., 2024). Digital twins provide a comprehensive picture of megacity operations by combining data from several sources, facilitating improved decision-making and advancing sustainability.

While digital twins offer significant opportunities to promote sustainable practices, their implementation faces several obstacles that need to be overcome. The high expense and complexity of creating and maintaining digital twins is one of the primary obstacles. This calls for the use of innovative technology like cloud computing, data analytics, and Internet of Things sensors, alongside the expertise needed for data management and analysis (Rasheed et al., 2023).

Data Security and Privacy

Due to the fact that digital twins require the acquisition and processing of substantial amounts of sensitive data, data security and confidentiality is crucial for businesses. Maintaining the safety and isolation of this data is vital in preventing unauthorized access and eventual misuse. This calls for implementing strong cybersecurity safeguards as well as adherence to relevant laws and standards. Notwithstanding these obstacles, digital twin provides enormous opportunities for advancing sustainability. Digital twin technology is becoming more accessible and cost-effective due to advancements in technology and increased data availability (Ammad et al., 2021). Similarly, the demand for more flexible architectures and a growing emphasis on sustainability are driving the adoption of comprehensive digital solutions across various industries.

Measuring Sustainability in the Digital Twin Context

In this context, sustainability refers to digital twins' capacity to improve social, economic, and environmental outcomes. To assess the effectiveness of digital binary operations in supporting sustainable practices, this extensive evaluation employs dynamic criteria and materials.

Key Metrics and Frameworks for Sustainability

Environmental Impact Assessment

Energy consumption is one of the main factors used to evaluate the sustainability of digital twins. Real-time energy use optimization and monitoring in multicolored systems are made possible by digital twins. Associations can cut carbon emigrations and functional costs by tracking patterns in energy consumption and implementing initiatives to reduce energy use. Digital twins encompass data from diverse sources and help in assessing the carbon footprint of operations (He et al., 2023). For associations hoping to achieve net-zero emigrations and nonsupervisory conditions, as well as sustainability pretensions, this data is crucial. Digital twins, for instance, can simulate the emigrations from transportation and artificial processes, providing a thorough understanding of the entire carbon footprint (Khan et al., 2024). Associations can use this knowledge to implement focused emigration reduction efforts, such as converting to sustainable energy sources or streamlining logistics (He et al., 2023).

Digital twins provide a thorough perception into material and resource overflows, lubricating the efficient use of funds. This statistic takes waste reduction and overall effectiveness into account by optimizing water operation, raw accessories, and other inputs. Construction directors can choose more sustainable decisions, cut waste, and maximize material use by analyzing this data (Zaland et al., 2024). Digital twins can also cover water and toxin operation in husbandry, ensuring optimal operation rates and reducing environmental impact.

Reducing the amount of waste generated is essential to sustainability. Real-time tracking and analysis of waste products by digital twins enables associations to implement trash reduction initiatives and enhance recycling and exercise protocols (Sengupta and Dreyer, 2023). For example, the kinds and quantities of waste produced during product manufacturing can be covered by the digital twin of a manufacturing unit. The installation can implement waste reduction strategies, such as streamlining product processes or recovering scrap accessories, by identifying trends and waste sources. This enhances resource efficiency and lowers disposal costs while promoting sustainability.

Life Cycle Assessment (LCA)

A thorough methodology called life cycle assessment, or LCA, is used to calculate how items and processes affect the environment from birth to death. Associations can obtain a more precise and comprehensive picture of their sustainability performance by combining LCA with digital binaries. The ability to evaluate the environmental effects of a product at each stage of its lifecycle from the procurement of raw materials to production, distribution, usage, and disposal—is improved by this integration.

Cradle-to-Grave Analysis

LCA entails evaluating a process's or product's environmental impact throughout the course of its lifecycle. By providing real-time data on every stage of the lifecycle, digital twin improves this analysis and makes it possible to identify environmental hotspots and provide more accurate assessments (El Bazi et al., 2023). For example, a digital twin of an electronic equipment can track resource consumption, and energy usage throughout its production, operational life, and disposal phases. This thorough examination aids in identifying phases that have major effects on the environment and implementing mitigation techniques.

Hotspot Identification

One of the key impacts of integrating LCA with digital twins is the ability to identify hotspots or locations with significant environmental impact (El Bazi et al., 2023). By identifying these areas, organizations can focus their efforts on implementing specific improvements, which will result in more sustainable practices. For example, a vehicle's digital twin can determine whether a certain product has a high carbon footprint. This intelligence enables producers to investigate necessary add-ons or procedures to lessen their environmental effect.

Continuous Improvement

The continuous monitoring and improvement of sustainability performance is made possible by the integration of digital technologies and LCA. Associations can enhance continuous progress towards sustainability pretensions by updating their plans and modernizing their LCA models when new data becomes available. In particular, a consumer electronics manufacturer can employ digital twins to continuously cover energy use and emigrations during product development. Through consistent integration of real-time data into their LCA models, they can discover novel opportunities to mitigate environmental impact and enhance sustainability performance.

Key Performance Indicators (KPIs)

Key Performance Indicators, or KPIs, are crucial instruments for gauging and comparing sustainability performance in the context of digital binary exchanges. KPIs can be tailored to particular activities and diligence, providing comparable and quantifiable data on many facets of sustainability. These guidelines assist associations in tracking their progress, identify areas needing improvement, and make well-informed decisions to improve their sustainability efforts.

Industry-Specific KPIs

Diverse diligence presents distinct sustainability opportunities and difficulties. To illustrate, KPIs for construction industry could be material waste, water operation, and energy effectiveness (Rajabi et al., 2022). KPIs in manufacturing may focus on controlling emigration, reducing waste, and improving product efficacy in the event of a disparity. KPIs that are adjusted to particular diligence will accurately represent the sustainability aspirations and precedents of every industry. KPIs in the transportation industry focus on energy effectiveness, emigrations per mile, and vehicle conservation efficacy; in the agrarian sector, they include crop production, soil health, and water usage effectiveness.

Measurable and Comparable Data

KPIs provide a consistent method for evaluating and contrasting sustainability performance among various associations and systems (Bilal et al., 2023). This network is essential for comparing fashionable practices and setting benchmarks. Associations can evaluate their progress towards sustainability pretensions and make data-driven decisions to improve performance by monitoring KPIs over time. A building company can analyze the energy efficiency of various structural systems using KPIs, linking techniques that improve performance and using those practices for future systems (Al Dakheel et al., 2020).

Challenges and Opportunities in Measuring Sustainability

Challenges

Data Availability and Quality: Maintaining data availability and quality is a major issue when utilizing digital tools to measure sustainability. Reliable and accurate data is necessary for insightful evaluations and analyses. However, gathering and combining data from disparate sources can be difficult and resource-intensive. To provide a complete picture of sustainability performance, for instance, data from IoT detectors, building operating systems, and external databases needs to be precise, harmonic, and compatible.

Standardization: Another difficulty is the absence of uniform materials and techniques for gauging sustainability performance (Bilal et al., 2022). Comparing performance across many systems and organizations can be tricky in the absence of defined criteria and grades. The development of fashionable behaviors and industry-wide norms is crucial for community and icing thickness in sustainability evaluations.

Cost and Complexity: Integrating sustainability frameworks with digital binary technology and implementing them can be both costly and complex (Alves et al., 2022). Besides developing the expertise required to manage and analyze the data, organizations must invest in advanced technologies like cloud computing, data analytics, and IoT detectors. However, these initial expenses may be offset by the long-term advantages in terms of sustainability and functional efficacy.

Opportunities

Technological Developments: Technological advancements have made digital twin technology more affordable and accessible. The barriers in implementing digital binary technology are coming down owing to advances in IoT detectors, data analytics, and cloud computing (Naeem et al., 2022). This will allow more associations to reap the benefits of this technology for sustainability (Saad et al., 2024). For example, improvements in detector technology are making it simpler to get precise real-time data on resource usage, emigration, and energy use.

Growing Emphasis on Sustainability: The shift away from digital binary technology is increasingly influenced by heightened emphasis on sustainability from consumers, regulators, and policymakers alike (Stuermer et al., 2017). Companies are increasingly recognizing the importance of digital transformation in surpassing competitors, building brand strength, and hitting sustainability goals. Businesses that thrive in sustainability, for instance, may be able to draw in more patrons and investors that value social and environmental responsibility.

Collaboration and Information Sharing: Encouraging cooperation and information sharing among government, academia, and industry can assist in addressing the difficulties involved in determining sustainability in the context of digital binary environments. Utilizing standardized materials and adopting best practices can help stakeholders enhance the effectiveness and sophistication of sustainability evaluations (Saad et al., 2024). Collaborative businesses can also promote the adoption of digital twins across various dynamic industries and raise awareness of their benefits.

Digital Twins and Green Building Certification

Green building certifications are recognized standards that support sustainable building techniques. These include BREEAM (Building Research Establishment Environmental Assessment Method) and LEED (Leadership in Energy and Environmental Design). Based on various sustainability criteria such as energy efficiency, water conservation, indoor environmental quality, and material selection, these tools assess buildings. Securing a green building certification not only demonstrates a dedication to sustainability but also provides a host of advantages, such as lower operating expenses, more occupant comfort, and improved marketability. Digital half, or virtual replicas of real-world structures, have come to light as crucial certifications for improving the green structure instrumentation process (Verma and Sangle, 2024). Digital twin technology provides precise and comprehensive data on the performance of structures, allowing engineers, architects, and building directors to maximize sustainability efforts from the design stage all the way through to operation and preservation as presented in Figure 3.

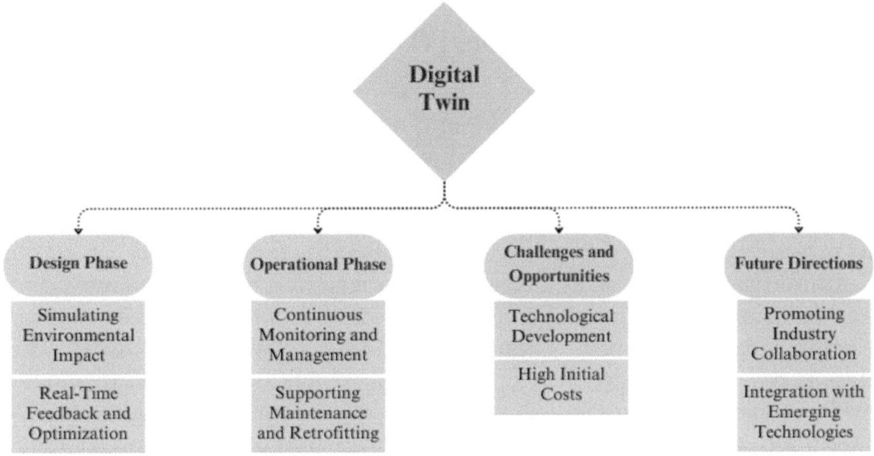

Figure 3: Digital Twins and green building certifications.

Digital Twins in the Design Phase

Digital twins enable engineers and masterminds to simulate dynamic design options throughout the design phase, allowing them to assess the opinions' potential environmental impact. The ability to choose the most environmentally friendly solutions that comply with green building instrument standards is essential. Aspects of structural design like exposure, choice of accessories can be modeled by digital twins.

Structure Exposure: A building's exposure has a big impact on how much energy it uses, especially for lighting, heating, and cooling. In order to help engineers place structures to maximize natural light and ventilation while minimizing heat gain or loss, digital twins can simulate the effects of various exposures on energy use. Specifically, energy demand for heating and cooling can be primarily decreased by orienting a structure to take advantage of unresisting solar heating during the day and natural shading during the summer.

Material Selection: A digital twin is able to calculate the energy footprint, continuity, and recyclability of accessories in addition to their overall environmental impact. This data-driven methodology guarantees that the selected accessories make a meaningful contribution to the overall sustainability of the structure. Digital twins, for instance, can assist in comparing the lifespan effects of conventional concrete with low-carbon concrete, influencing perceptions towards more environmentally friendly solutions.

Digital Twins in the Operational Phase

Imagine a smart office block that aims to obtain LEED Platinum certification. Using a digital twin, engineers are able to simulate vibrant design choices, optimal structural exposure, accessory selection, and natural light and ventilation methods during the design phase. Real-time input is provided by the digital binary, which enables the design crew to make data-driven decisions that improve the sustainability profile of the building. Real-time data on water operation, energy consumption, and interior environmental quality is provided via IoT devices that are strategically placed throughout the building (Haris et al., 2024). This data is broken down by the digital binary, which identified inefficiencies and recommended remedial action. As an example, the digital binary identified that the HVAC system has been operating during off-peak hours without necessity and suggested adjusting the schedule, resulting in notable energy savings.

Supporting Maintenance and Retrofitting

Digital twins are invaluable not only in the design and functioning stages but also in the preservation and retrofitting of actual constructions. Digital twins deliver a holistic perspective of the performance of the structure, which can help inspire forward-thinking conservation plans and pinpoint opportunities for upgrading to improve sustainability (Cinquepalmi and Cumo, 2022).

In order to preserve and improve sustainability performance, retrofitting is required as structures get older. The ability of digital twins to simulate several build scripts aids structure directors in determining the most economical and long-lasting outcomes. It could simulate the effects of modernizing sequestration, setting up solar panels, or swapping out old HVAC systems for more efficient ones. These models guarantee that constructed systems yield the greatest possible benefits for sustainability and assist in achieving sophisticated scenarios for green building instruments. For example, a digital twin can simulate the effects of adding solar panels to the roof, assisting in the selection of the best location and size for generating the most energy.

Navigating Challenges and Opportunities

Implementing digital twins to enhance sustainability poses several challenges despite their potential benefits. It is essential to comprehend and tackle these obstacles in order to fully reap the rewards of digital technologies and ensure their seamless integration throughout diverse fields. The opportunities for improving sustainability through digital twins are enormous, notwithstanding the challenges.

Challenges in Implementing Digital Twins

Technological Requirements

A large investment in cutting-edge technologies is necessary for the development and upkeep of the digital half. Important components include.

IoT Sensors: In order to create and streamline digital twins, these detectors gather data in real time from physical systems. IoT sensors conservation and installation can be costly, especially for large-scale businesses or intricate systems.

Data analytics: Sophisticated data analytics tools are required to process and analyze the enormous amounts of data provided by IoT sensors (Plageras et al., 2018). These instruments require a significant amount of processing power and specialized software in order to extract practical perceptivity from raw data.

Expertise: Digital twin implementation and management call for a high degree of skill. This encompasses domain-specific expertise, software engineering abilities, and data science abilities (Ali et al., 2024). Hiring and keeping qualified workers can be costly and difficult.

Integration and Interoperability

Digital twins need to work seamlessly with current workflows and systems. This integration may be difficult because of:

Compatibility Problems: It can be challenging to guarantee compatibility between new digital twin technologies and legacy systems. This frequently calls for either specialized solutions or significant adjustments to current systems.

Data Integration: Merging information from multiple sources, including external databases, corporate systems, and Internet of Things sensors, presents a number of difficulties. For a complete picture of the system being modeled, the data must be reliable, consistent, and compatible.

Data Security and Privacy

As digital systems need the collection and processing of vast amounts of sensitive information, data security and sequestration are essential for businesses implementing digital divides. Preventing unauthorized access and implicit abuse requires improving the security and sequestration of this data. Key considerations include:

Strong Cybersecurity Implementations: To protect data from cyber hazards, strict cybersecurity regulations are essential. To safeguard data integrity and confidentiality, it involves employing encryption, firewalls, intrusion detection systems, and regular security audits.

Regulation Compliance: Similar to the Californian Customer Protection Act (CCPA) in the US and the GDPR (General Data Protection Regulation) in Europe as a whole, companies must comply with and adhere to the relevant confidentiality regulations and norms. By ensuring that data is handled morally, sensibly, and with regard for individuality, compliant safeguards such as the right to privacy. Strict access controls must be implemented in order to guarantee that only workers with authorization can access sensitive information. Using multi-factor authentication, premise-based access controls, and routine access log monitoring are all examples of this.

Opportunities by Digital Twins

Despite these challenges, the opportunities obtainable by digital twins for driving sustainability are immense. Advances in technology and the adding of data are making digital twins more accessible and cost-effective.

Improved Data Analytics: Advances in machine learning and data analytics are making it easier to analyze and repurpose big databases, providing more insight and precise forecasts. With the use of these technologies, organizations can improve sustainability performance, cut waste, and streamline operations.

Cloud Computing: This method of storing and recycling the enormous amounts of data produced by digital twins provides scalable and economic outcomes. Cloud platforms make digital twins more accessible to associations of all sizes by providing the processing power and storage capacity required to execute complicated simulations and analyses (Zhang et al., 2022).

Regulatory Assistance: Encouraging organizations to use digital twins is the addition of nonsupervisory assistance for sustainable practices. The provision of incentives, subsidies, and subventions by governments and nonsupervisory organizations for the implementation of sustainable technology is making it more financially feasible for associations to invest in digital half twins.

Customer Demand: As consumers become more aware of environmental issues, they favor goods and services that are produced responsibly (Rasheed et al., 2023). Businesses that adapt sustainability by influencing digital twins can improve their brand image and draw in eco-aware customers.

Future Research and Development Efforts

Subsequent research and development endeavors must concentrate on tackling the obstacles linked with the execution of digital twins and investigating novel applications that promote sustainability. Important areas of attention consist of,

Developing Standardized Frameworks and Methodologies

For digital twins to be widely relinquished and effective, formalized materials and procedures for monitoring and evaluating their environmental performance are essential. These materials ought to provide precise instructions for gathering, analyzing, and reporting data so that organizations can standardize their operations and pinpoint fashionable behaviors.

Industry Cooperation: Standards organizations, nonsupervisory bodies, and industry colleges ought to collaborate to create and disseminate sophisticated procedures and uniform guidelines for digital binary offenses. Through this partnership, interoperability problems can be resolved and it will be ensured that digital components can smoothly merge with physical systems.

Academic Research: To develop novel approaches and instruments for evaluating the sustainability impact of digital twins, academic institutions should carry out research. This investigation will provide invaluable insight and aid in the creation of standardized materials.

Promoting Collaboration Between Industry, Academia, and Government

Encouraging cooperation between government, academia, and diligence is crucial to expanding the use of digital twins in the sustainability dimension. Stakeholders can share expertise, produce creative solutions, and tackle shared issues by cooperating.

Public-Private Partnerships: These collaborations have the potential to facilitate the advancement and application of digital binary technology. Private businesses can provide specialized funds, while governments can offer backing and nonsupervisory help.

Participating in knowledge Platforms: Creating knowledge-sharing venues that resemble conventions, stores, and internet discussion boards can facilitate the sharing of concepts, fashionable methods, and completed assignments among interested parties. These platforms have the potential to stimulate innovation in digital binary technology and create a cooperative environment.

Exploring New Applications for Driving Sustainability

Agriculture and Food Production: By maximizing resource utilization, addressing crop health, and refining force chain performance, digital twins can improve the sustainability of husbandry and food products. For instance, digital twins can simulate how various irrigation techniques affect crop output and water usage, assisting farmers in adopting more environmentally friendly techniques.

Renewable Energy: Digital twins can facilitate the grid's integration of renewable energy sources, like wind and solar power. Digital twins can optimize the placement, operation, and conservation of renewable energy systems, hence promoting a more sustainable energy system, by modeling the installation's performance and impact (Rasheed et al., 2024).

To maximize the potential and promote sustainability, digital twins must be navigated through the opportunities and difficulties they present. Technological advancements and the increased attention on sustainability present enormous opportunities for employing digital twins, despite the high cost, complexity, and data security concerns presenting considerable challenges. Future research and development efforts should focus on resolving these issues, creating standardized materials, and investigating novel approaches to guarantee that digital technologies can successfully support sustainable practices throughout diverse fields of endeavor.

References

Al Dakheel, J., Del Pero, C., Aste, N. and Leonforte, F. (2020). Smart buildings features and key performance indicators: A review. *Sustainable Cities and Society*, 61: 102328. https://doi.org/10.1016/j.scs.2020.102328.

Ali, Z., Saad, S., Rasheed, K. and Ammad, S. (2024). AI future perspectives and trends in construction. In *AI in Material Science* (pp. 239–261). CRC Press. https://doi.org/10.1201/9781003438489-11.

Alves, L., Ferreira Cruz, E., Lopes, S. I., Faria, P. M., Rosado da Cruz, A. M. et al. (2022). Towards circular economy in the textiles and clothing value chain through blockchain technology and IoT: A review. *Waste Management & Research: The Journal for a Sustainable Circular Economy*, 40(1): 3–23. https://doi.org/10.1177/0734242X211052858.

Ammad, S., Saad, S., Bashir, M. T., Qureshi, A. H., Altaf, M. et al. (2021). Construction accidents via integrating Building Information Modelling (BIM) with emerging digital technologies a review. *2021 3rd International Sustainability and Resilience Conference: Climate Change*. https://doi.org/10.1109/IEEECONF53624.2021.9668066.

Annaram, A. R. (2022). Optimizing physician lifecycle management: Streamlining processes with BPM tools in healthcare. *Journal of Health Statistics Reports*, 1–3. https://doi.org/10.47363/JHSR/2022(1)116.

Attaran, M. and Celik, B. G. (2023). Digital Twin: Benefits, use cases, challenges, and opportunities. *Decision Analytics Journal*, 6(November 2022): 100165. https://doi.org/10.1016/j.dajour.2023.100165.

Badshah, I., Saad, S., Ammad, S., Rasheed, K., Khan, M. W. et al. (2024). Enhancing sustainable mobility: Lane change strategies for efficient traffic flow at stationary bottlenecks. *2024*

ASU International Conference in Emerging Technologies for Sustainability and Intelligent Systems (ICETSIS), 1197–1202. https://doi.org/10.1109/ICETSIS61505.2024.10459583.

Bécue, A., Maia, E., Feeken, L., Borchers, P., Praça, I. et al. (2020). A new concept of Digital Twin supporting optimization and resilience of factories of the future. *Applied Sciences*, 10(13): 4482. https://doi.org/10.3390/app10134482.

Bilal, M., Iftikhar Ali, M., Rasheed, K. and Saad, S. (2023). Exploring adoption trends of knowledge management systems in the software industry's SMEs: A comprehensive investigation of enablers and barriers. *2023 4th International Conference on Data Analytics for Business and Industry (ICDABI)*, 598–604. https://doi.org/10.1109/ICDABI60145.2023.10629373.

Bilal, M., Rasheed, K., Abbasi, M. F., Yamin, I., Khan, B. S. A. et al. (2022). Impact of business strategy on project management elements focus moderating role of competition attributes in textile industry. *2022 International Conference on Decision Aid Sciences and Applications (DASA)*, 719–724. https://doi.org/10.1109/DASA54658.2022.9765222.

Bofill, J., Abisado, M., Villaverde, J. and Sampedro, G. A. (2023). Exploring Digital Twin-based fault monitoring: Challenges and opportunities. *Sensors*, 23(16): 7087. https://doi.org/10.3390/s23167087.

Cinquepalmi, F. and Cumo, F. (2022). Using Digital Twin Models (DTM) for managing, protecting and restoring historical buildings. *Conservation Science in Cultural Heritage*, 22(1 SE-Articles): 425–445. https://doi.org/10.6092/issn.1973-9494/17364.

El Bazi, N., Mabrouki, M., Laayati, O., Ouhabi, N., El Hadraoui, H. et al. (2023). Generic multi-layered digital-twin-framework-enabled asset lifecycle management for the sustainable mining industry. *Sustainability*, 15(4): 3470. https://doi.org/10.3390/su15043470.

Elgazzar, K., Khalil, H., Alghamdi, T., Badr, A., Abdelkader, G. et al. (2022). Revisiting the Internet of Things: New trends, opportunities and grand challenges. *Frontiers in the Internet of Things*, 1. https://doi.org/10.3389/friot.2022.1073780.

Franciosi, C., Miranda, S., Veneroso, C. R. and Riemma, S. (2022). Improving industrial sustainability by the use of digital twin models in maintenance and production activities. *IFAC-PapersOnLine*, 55(19): 37–42. https://doi.org/10.1016/j.ifacol.2022.09.215.

Haris, M., Saad, S., Ammad, S. and Rasheed, K. (2024). AI in fabrication and construction. In *AI in Material Science* (pp. 169–192). CRC Press. https://doi.org/10.1201/9781003438489-8.

Haris, M., Saad, S., Ammad, S., Rasheed, K., Aman, H. et al. (2024). Integrating sustainable practices for effective solid waste management in peshawar: Overcoming challenges and promoting environmental. *2024 ASU International Conference in Emerging Technologies for Sustainability and Intelligent Systems (ICETSIS)*, 1–7. https://doi.org/10.1109/ICETSIS61505.2024.10459445.

He, B. and Mao, H. (2023). Digital Twin-Driven product sustainable design for low carbon footprint. *Journal of Computing and Information Science in Engineering*, 23(6). https://doi.org/10.1115/1.4062427.

He, B., Mao, H., Li, T. and Xiao, J. (2023). A closed-loop digital twin modeling method integrated with carbon footprint analysis. *Computers & Industrial Engineering*, 182: 109389. https://doi.org/10.1016/j.cie.2023.109389.

Hu, L. (2023). Research on the application of Digital Twin in smart cities. *Advances in Economics, Management and Political Sciences*, 42(1): 14–20. https://doi.org/10.54254/2754-1169/42/20232072.

Jafari, M., Kavousi-Fard, A., Chen, T. and Karimi, M. (2023). A review on Digital Twin technology in smart grid, transportation system and smart city: Challenges and future. *IEEE Access*, 11: 17471–17484. https://doi.org/10.1109/ACCESS.2023.3241588.

Katsoulakis, E., Wang, Q., Wu, H., Shahriyari, L., Fletcher, R. et al. (2024). Digital twins for health: A scoping review. *NPJ Digital Medicine*, 7(1): 77. https://doi.org/10.1038/s41746-024-01073-0.

Khan, M. I., Khan, N., Saad, S. and Khan, M. W. (2024). Artificial Intelligence in construction materials. In *AI in Material Science* (pp. 47–64). CRC Press. https://doi.org/10.1201/9781003438489-3.

Leng, J., Wang, D., Shen, W., Li, X., Liu, Q. et al. (2021). Digital twins-based smart manufacturing system design in Industry 4.0: A review. *Journal of Manufacturing Systems*, 60: 119–137. https://doi.org/10.1016/j.jmsy.2021.05.011.

Naeem, U. H., Bilal, M., Syed, F., Rasheed, K., Saad, S. et al. (2022). A multilayer encryption model to protect healthcare data in cloud environment. *2022 International Conference on Data Analytics for Business and Industry, ICDABI 2022*. https://doi.org/10.1109/ICDABI56818.2022.10041708.

Nativi, S., Mazzetti, P. and Craglia, M. (2021). Digital ecosystems for developing Digital Twins of the earth: The destination earth case. *Remote Sensing*, 13(11): 2119. https://doi.org/10.3390/rs13112119.

Negri, E., Fumagalli, L. and Macchi, M. (2017). A review of the roles of Digital Twin in CPS-based production systems. *Procedia Manufacturing*, 11: 939–948. https://doi.org/10.1016/j.promfg.2017.07.198.

Papanagnou, C. I. (2020). *A Digital Twin Model for Enhancing Performance Measurement in Assembly Lines* (pp. 53–66). https://doi.org/10.1007/978-3-030-18732-3_4.

Plageras, A. P., Psannis, K. E., Stergiou, C., Wang, H., Gupta, B. B. et al. (2018). Efficient IoT-based sensor BIG Data collection–processing and analysis in smart buildings. *Future Generation Computer Systems*, 82: 349–357. https://doi.org/10.1016/j.future.2017.09.082.

Rajabi, S., El-Sayegh, S. and Romdhane, L. (2022). Identification and assessment of sustainability performance indicators for construction projects. *Environmental and Sustainability Indicators*, 15: 100193. https://doi.org/10.1016/j.indic.2022.100193.

Rasheed, K., Saad, S., Ammad, S. and Badshah, I. (2023). IoT enhanced solar device management: Integrating remote surveillance and energy efficiency for sustainable operation. *2023 TRON Symposium (TRONSHOW)*, 1–6.

Rasheed, K., Saad, S., Ammad, S. and Bilal, M. (2023). Integrating sustainability management and lean practices for enhanced supply chain performance: Exploring the role of process optimization in SMEs. *Engineering Proceedings*, 56(1). https://doi.org/10.3390/ASEC2023-16370.

Rasheed, K., Saad, S., Ammad, S., Iftikhar Ali, M., Bilal, M. et al. (2023). Towards sustainable growth management: Analyzing the consequences of ecological deterioration in asian countries on energy consumption, economic growth and financial progress. *2023 International Conference on Sustainable Islamic Business and Finance (SIBF)*, 31–36. https://doi.org/10.1109/SIBF60067.2023.10379892.

Rasheed, K., Waqas, M., Saad, S., Zaland, A., Ali, Z. et al. (2024). Ethical and societal implications of AI in construction. In *AI in Material Science* (pp. 216–238). CRC Press. https://doi.org/10.1201/9781003438489-10.

Saad, S., Alaloul, W. S., Rasheed, K. and Ammad, S. (2022). Modelling and simulation of construction cyber-physical systems. In *Cyber-Physical Systems in the Construction Sector* (1st ed., pp. 88–110). CRC Press.

Saad, S., Haris, M., Ammad, S. and Rasheed, K. (2024). AI-assisted building design. In *AI in Material Science* (pp. 143–168). CRC Press. https://doi.org/10.1201/9781003438489-7.

Saad, S., Rasheed, K., Ammad, S., Waqas Khan, M., Zaland, A. et al. (2024). Is AI the architect of tomorrow's materials in the age of Industry 4.0? In *AI in Material Science* (pp. 1–14). CRC Press. https://doi.org/10.1201/9781003438489-1.

Saad, S., Zaland, A., Ammad, S. and Rasheed, K. (2024). Robotics in construction. In *AI in Material Science* (pp. 115–142). CRC Press. https://doi.org/10.1201/9781003438489-6.

Schleich, B., Anwer, N., Mathieu, L. and Wartzack, S. (2017). Shaping the digital twin for design and production engineering. *CIRP Annals*, 66(1): 141–144. https://doi.org/10.1016/j.cirp.2017.04.040.

Sengupta, S. and Dreyer, H. (2023). Realizing zero-waste value chains through digital twin-driven S&OP: A case of grocery retail. *Computers in Industry*, 148: 103890. https://doi.org/10.1016/j.compind.2023.103890.

Soori, M., Arezoo, B. and Dastres, R. (2023). Digital twin for smart manufacturing, A review. *Sustainable Manufacturing and Service Economics*, 2, 100017. https://doi.org/10.1016/j.smse.2023.100017.

Stuermer, M., Abu-Tayeh, G. and Myrach, T. (2017). Digital sustainability: Basic conditions for sustainable digital artifacts and their ecosystems. *Sustainability Science*, 12(2): 247–262. https://doi.org/10.1007/s11625-016-0412-2.

Tao, F., Zhang, H., Liu, A. and Nee, A. Y. C. (2019). Digital Twin in Industry: State-of-the-Art. *IEEE Transactions on Industrial Informatics*, 15(4): 2405–2415. https://doi.org/10.1109/TII.2018.2873186.

Verma, P. and Sangle, P. (2024). Role of digital transformation in inspection and certification. In *Handbook of Quality System, Accreditation and Conformity Assessment* (pp. 1–29). Springer Nature Singapore. https://doi.org/10.1007/978-981-99-4637-2_28-1.

Waqas Khan, M., Saad, S., Ammad, S., Rasheed, K., Jamal, Q. et al. (2024). Smart infrastructure and AI. In *AI in Material Science* (pp. 193–215). CRC Press. https://doi.org/10.1201/9781003438489-9.

Wook, T. and Mo, Y. (2024). Developments in the built environment a comprehensive digital twin framework for building environment monitoring with emphasis on real-time data connectivity and predictability. *Developments in the Built Environment*, 17(December 2023): 100309. https://doi.org/10.1016/j.dibe.2023.100309.

Yang, B., Lv, Z. and Wang, F. (2022). Digital Twins for intelligent green buildings. *Buildings*, 12(6): 856. https://doi.org/10.3390/buildings12060856.

Yu, W., Patros, P., Young, B., Klinac, E., Walmsley, T. G. et al. (2022). Energy digital twin technology for industrial energy management: Classification, challenges and future. *Renewable and Sustainable Energy Reviews*, 161, 112407. https://doi.org/10.1016/j.rser.2022.112407.

Zaland, A., Saad, S., Rasheed, K. and Ammad, S. (2024). Automation in construction industry. In *AI in Material Science* (pp. 88–114). CRC Press. https://doi.org/10.1201/9781003438489-5.

Zhang, G., MacCarthy, B. L. and Ivanov, D. (2022). The cloud, platforms, and digital twins—Enablers of the digital supply chain. In *The Digital Supply Chain* (pp. 77–91). Elsevier. https://doi.org/10.1016/B978-0-323-91614-1.00005-8.

Zhang, Y. Z. (2018). Virtual reality technology. In *Digital Orthopedics* (pp. 21–33). Springer Netherlands. https://doi.org/10.1007/978-94-024-1076-1_3.

Zhang, Z., Wei, Z., Court, S., Yang, L., Wang, S. et al. (2024). A review of Digital Twin technologies for enhanced sustainability in the construction industry. *Buildings*, 14(4): 1113. https://doi.org/10.3390/buildings14041113.

Zhong, D., Xia, Z., Zhu, Y. and Duan, J. (2023). Overview of predictive maintenance based on digital twin technology. *Heliyon*, 9(4): e14534. https://doi.org/10.1016/j.heliyon.2023.e14534.

Chapter 6

Digital Twin and Circular Economy in Construction

Ahmad Zaland,[1] *Kumeel Rasheed,*[1,*] *Syed Saad,*[2]
Syed Ammad[3] and *Ali Rostami*[4]

Introduction

Natural resource use has historically taken a traditional linear approach, predicated on the idea that resources are acquired, used, and then disposed of as waste. Construction is one of the main industries that is progressively moving from the linear to a circular economy. Construction's adoption of the circular economy is moving slowly because of a number of obstacles and difficulties. Construction can overcome these obstacles with the aid of digital technologies. The digital twin (DT), a new generation of digital technologies, is now rarely ever utilized in the building of the circular economy (Meng et al., 2023). The sustainability of the natural environment was seriously challenged by this "take-make-dispose" strategy, which led to the squandering of natural resources and the generation of carbon dioxide. 35% of carbon dioxide emissions and 50% of raw material consumption are attributable to construction, a key industry sector in many nations (Numanovich and Abbosxonovich, 2020).

[1] Department of Civil Engineering, CECOS University of IT and Emerging Sciences, Pakistan.
[2] Science, Engineering & Manufacturing Centre Sandwell College West Bromwich, United Kingdom.
[3] School of Engineering and Built Environment, Griffith University, Gold Coast, Australia.
[4] Department of Civil Engineering and Built Environment, Liverpool John Moores University, Liverpool, United Kingdom.
* Corresponding author: kumeelrasheed@cecosian.edu.pk

There is a request for a paradigm change in the construction sector from the linear economy to the circular economy in order to decrease waste from natural resources, reduce carbon dioxide emissions, and ultimately safeguard the environment. By utilizing by-product exchange, product reuse, and material recycling products as foundational elements, the circular economy accomplishes its strategic goals (Hossain et al., 2020). Consequently, the shift from the "cradle to grave" construction model to the "cradle to cradle" framework is often termed an industrial revolution. The circular economy is necessary, but there are a number of obstacles or difficulties in accomplishing its objectives in the building industry. Organizational, economical, technical, social, political, and environmental barriers are the six categories into which circular economy constraints in construction are categorized based on an extensive literature analysis (Patwa et al., 2021). Eleven categories of barriers related to the circular economy are discussed in relation to building, namely: supply chain, stakeholder, technical, organizational, cultural, market, knowledge, financial, managerial, and regulatory.

A virtual model that serves as a copy of the physical asset and gathers and transmits real-time data using technologies like sensors, communication networks, and 3D models is known as a "digital twin." The digital twin holds the ability to more effectively gather, organize, and present circular economy information when compared to other digital technologies. It is therefore feasible that the digital twin will contribute more to the development of the circular economy (Chauhan et al., 2022). While the digital twin can serve several purposes during the various stages of a construction project, including design, construction, operation and maintenance, and destruction, its application in the context of the circular economy has not received much attention in the field of construction. A theoretical approach to utilizing the digital twin for the circular economy in construction is one of the few relevant researches available.

Defining Digital Twin within Circular Economy Principles

The necessity of sustainability has gained a lot of attention lately and is now a crucial goal on political agendas both in Europe and around the world. Politicians have developed the idea of a resource-efficient and greenhouse gas-neutral economy via the United Nations Sustainable Development Goals and the European Green Deal (Fallah Shayan et al., 2022). Making the transition from a linear to a circular economy is one way to assist politicians in the area of more resource-efficient value creation. Extending a product's lifetime and maintaining materials and components in the cycle are two strategies used by the circular economy to achieve the goal of resource conservation (Rasheed et al., 2024). However, the involvement of numerous stakeholders and varied information systems, which impede coordination and transparency, make information exchange a significant difficulty. Therefore, efficient information sharing is essential to guaranteeing the effectiveness and prosperity of a circular economy (Milios, 2021). This makes sure that everyone who needs to know about a product or material cycle has the knowledge they need to make decisions that support the circular economy. Information transfer is disrupted the instant a product is given to the end user, even if it often functions very well until the product is put to use. Furthermore, information on how the product is used, information on maintenance and repair needs, and projections for when its useful life would expire are frequently lacking. However, in order to perform repairs or determine whether a component is appropriate for reuse, repair service providers, remanufacturers, or dismantlers need information about the product's state.

Digital solutions that process pertinent information about individual assets for relevant stakeholders at the appropriate moment during the product lifecycle (PLC), such as the digital twin, can support a data-driven implementation of CE (Mügge et al., 2024). A viable strategy for creating a data-driven ecosystem that supports new businesses and value creation models for all business sizes, from small medium enterprises (SMEs) to major corporations, is the idea of a "digital twin" for the circular economy. Numerous international research initiatives, including the German projects RECLAIM and Catena-X, are now investigating this problem (Mügge et al., 2023).

Nevertheless, no comprehensive analysis of the digital twin's readiness and actual potential for the circular economy exists. This chapter identifies, from a holistic viewpoint, the possible and existing sustainability-oriented scenarios for the implementation of digital twins for circular economy across the PLC and assesses them according to their technology readiness level (TRL).

Features of Current Technologies for Information Systems

Cross-product life cycle information management is challenging due to differences in today's information systems. When a product is turned over to the end user, there is typically a lapse in information, even though the initial stages of a PLC perceive a fair amount of information gathering, acquisition, and interchange. When a product enters its use phase, it is frequently the case that no additional information about it is noted or documented and cannot be shared as a result (Haris et al., 2024). A product life cycle involves a range of actors with disparate information systems, which makes it challenging to exchange current information without appropriate interfaces and information transformation into interchangeable data formats (Ford et al., 2013). The geometric and conceptual diversity of the various information systems presents additional difficulties in addition to the technical heterogeneity. Differences in the way that the same facts are structured (structural heterogeneity), how the same terms are interpreted (semantic heterogeneity), and the backgrounds of the various players' knowledge can all contribute to misunderstandings. Furthermore, when gathering, transferring, and utilizing information, organizational and regulatory requirements for data security and protection must be taken into account. Notwithstanding the technological, structural, and semantic difficulties, stakeholders are frequently reluctant to exchange information with one another if they do not perceive a clear personal benefit (Saad et al., 2024). Because of this, information that is crucial to the effective distribution of goods and commodities is frequently unavailable. Within the circular economy framework, the concepts of digitization and digital twins hold significant untapped potential for further research. Concepts related to digitization have the ability to improve information management and increase transparency about pertinent data for the distribution of goods and commodities (Moghrabi et al., 2023).

Use Cases for Digital Twins in the Circular Economy

By supporting the circular economy, establishing the required transparency across the PLC, and guaranteeing a closed-loop system at the product, components material, process, and data levels, a digital twin for it aims to address this predicament. Digital twins provide the ongoing gathering, analysis, and assessment of data, giving different stakeholders access to this information over the course of the project. New services and business models are arising as a result. Figure 1 illustrates the discoveries for the

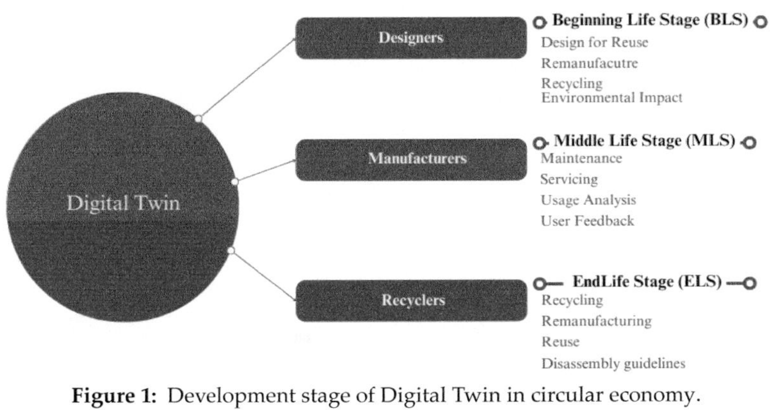

Figure 1: Development stage of Digital Twin in circular economy.

concentrated development stage for digital twin applications in a circular economy.

Circular Economy for the Beginning Life Stage with Digital Twins

The possibility to use data from the entire PLC to optimize product design early in the product development process, including production, development, and product planning, is encompassed in the Beginning Life Stage (BLS) (Lewis et al., 2017). Conclusions about design needs pertaining to longevity and user behavior can be drawn from the data from the digital twin throughout the PLC (Feedback to Design). Product design can be optimized for better reuse, remanufacturing, or recycling by taking end of life (ELS) treatment and regional conditions into consideration (Design for Reuse, Remanufacture, Recycling). Products with a longer life can be created by locating, examining, and assessing new product configurations (Joshi and Gupta, 2019). Considering all factors, enhanced product design techniques (Design for Environment) can be utilized to monitor and reduce environmental impact.

Additionally, taking into account the idea of ecological design which includes material compatibility and selection can help support decisions made during the product design process in terms of sustainability (Saad et al., 2024). This entails evaluating circular potential and accomplishments, tracking waste, emissions, and discharges in real-time, and analyzing and monitoring material inputs and outputs. It is possible to quantify greenhouse gas emissions using these methods (Vezzoli, 2018). Furthermore, the digital twin for the circular economy has the ability to forecast particular metrics, like future CO_2 emissions or recycling rates. Important choices on life cycle assessments at BLS can be made using the information displayed in a digital twin dashboard (Birat, 2015). The selection of suppliers and transportation routes can also have an impact on sustainability. Additionally, by keeping an eye on the system's condition and energy usage, the manufacturing process' sustainability can be increased. In addition, it is possible to prevent needless downtime and maintain the production system in accordance with specifications. It also emphasizes how many more laws there are that need to be taken into account when developing new products (design for compliance). The majority of these rules seek to minimize hazardous materials, improve product reparability, lengthen product lifespans, and guarantee that energy efficiency standards are met.

Circular Economy for the Middle Life Stage with Digital Twins

Service, use, and distribution are all included in the Middle Life Stage (MLS). According to multiple sources, a digital twin for circular economy can facilitate extended product use by analyzing condition data based on user behavior and providing information about required maintenance and servicing. In addition to technical features for maintenance, the digital twin for the circular economy can provide information about the state of the product. As such, this capacity allows for the simultaneous improvement of driving comfort and safety and the reduction of fuel consumption, especially in the automotive sector. Furthermore, learnings from product usage provide insightful direction for creating new product versions in the future (Eladly et al., 2023). In the sharing economy, location data facilitates a demand-driven distribution of shared products within designated zones, while product usage insights allow tailored pricing for consumers.

Circular Economy for the End Life Stage with Digital Twins

End Life Stage (ELS) encompasses recovery, recycling, remanufacturing, and reuse. According to the pertinent research, the digital twin for circular economy at the ELS can assist ELS decision-making regarding the best course of action for recycling materials or reusing or remanufacturing parts by giving the required details about the product's past and present conditions. This information might contain, for instance, how long products, parts, and materials are predicted to last. In ideal circumstances, the asset condition manual inspection procedure can be greatly accelerated or even skipped entirely if the digital twin has pertinent information that would normally need laborious manual inspections. In this phase, digital twins also aid in the virtual product disassembly, lowering product complexity through information provision, and enhancing the effectiveness of disassembly procedures. In addition, guidelines for recycling and disassembly can be supplied (Wang and Wang, 2019).

Digital Twin as a Catalyst for Circular Construction Practices

It is widely understood that there is a problem in expediting sustainable changes to guarantee the survival of the human population, the environment, and the global economy. According to the European Green Deal, the European Union member states have set ambitious goals to reduce emissions by 55% from 1990 levels by 2030 and transition to a society that is climate neutral, decoupling economic growth from resource depletion and environmental deterioration (den Elzen et al., 2022). The building industry is one important area that is essential to completing this shift. In the Netherlands, building and maintaining homes, workplaces, roads, and other infrastructure accounts for the third-largest resource footprint (32 million tonnes), but it also consumes the most raw materials overall (nearly 29 million tonnes), according to the most recent circularity gap report (de Wit et al., 2020). To date, two primary approaches have been proposed in scenarios to establish a more circular construction economy: zero demolition and significant modifications to the way buildings are constructed.

Current Accomplishments Include Information Sharing, Pilots and Models, and Documentation

Buildings as Material Banks (BAMB), a major initiative supported by the European Union, is one of the main efforts to drastically convert the construction industry in Europe into a circular economy. The goal of BAMB is to alter the perception of building materials in order to encourage more material reuse and, consequently, less waste and dependency on virgin resources. BAMB worked with a stakeholder network of building material and installation producers/suppliers, construction companies, builders, real estate and construction owners, resource managers, architects, engineers, logistics managers, real estate consultants, recycling and deconstruction companies, policy makers, and researchers to develop pilots that showed the viability of systemic change in the industry. This network included 15 partners from knowledge institutes, consultants, contractors, and municipalities. In the end, four pilot projects were completed: the Build Reversible in Conception (BRIC) wooden building, constructed by young trainees; the Circular Retrofit Lab (CRL), which renovates prefabricated student housing modules for various uses; the Reversible Experience Modules (REMS), a flexible indoor interactive and modular exhibition space; and a steel-framed building module in the Green Transformable Building (GTB) Lab.

The pilots in BAMB are guided by three concepts. First, deconstruction-friendly design makes it possible to install materials so they can be later taken apart and put back together in a different structure or area. By implementing standardization and dry connections in conjunction with prefabricating reconfigurable modules, design for deconstruction can enable the reuse of components and modules, thereby prolonging the material's usable life. Second, information about the kind, configuration, volume, and location of materials can be recorded and shared throughout the value chain by building a database of materials known as materials passports. This could lead to the development of a market for the reuse of building materials. Third, circular business models can promote the reuse of building materials, components, and modules by moving the emphasis from a product-dominant logic to a service-dominant logic. Typical strategies designed to shift the emphasis from ownership to facilitating reuse by granting access to building materials are take-back and leasing programmes.

Even after these pilots have demonstrated their capabilities, a number of important issues still need to be resolved. It is important to remember that design for deconstruction is still in its early phases of development. Currently, less than 1% of structures are completely removable. While designing for demolition can be a practical approach to developing new structures, dealing with the stock of existing buildings still presents a barrier. In fact, off the four pilots in BAMB, only CRL addressed the context of renovation; the other three focused on new constructions. The fact that materials in existing buildings are frequently in composite form, where supply chains and logistical processes to disaggregate materials and components are still in their infancy, can make attempts to reuse and recycle materials in building renovations even more challenging. This creates additional obstacles to guaranteeing the quality of reused materials (Oliveira et al., 2024).

Optimizing Designs for Circularity

Circular construction techniques are centered on design optimization, and digital twins are essential in helping stakeholders integrate circularity into a building or infrastructure projects from the very beginning (Çetin, 2023). Here, we dive into the complexities of design optimization in the context of circularity, examining how digital twins help create structures that are not just practical and visually beautiful but also resource-efficient, sustainable, and adaptive.

Algorithms for Generating Parametric Designs

By utilizing generative algorithms and parametric modeling, digital twins enable designers, architects, and engineers to investigate an extensive range of design options. With the aid of these computational design tools, stakeholders can produce and assess design choices iteratively in accordance with predetermined performance objectives and restrictions. Design characteristics like material selection, modularity, and flexibility can be integrated with circularity concepts to enable stakeholders optimize designs for recycling, reuse, and disassembly right from the start.

Selection of Materials and Lifecycle Evaluation

The careful selection of materials with consideration for their end-of-life disposal and environmental impact is fundamental to circular design. The integration of databases and lifecycle assessment (LCA) technologies is made easier by digital twins, allowing stakeholders to assess the environmental impact of materials at every stage of their lifecycle from extraction and manufacture to use and disposal (Bilal et al., 2023). Through a thorough analysis of the energy, resource, and embodied carbon requirements of various material possibilities, designers may make well-informed decisions that give priority to materials with lower environmental impact and higher potential for circularity (Boje et al., 2023).

Disassembleable and Modular Architecture

Digital twins make it easier to build structures that are naturally demountable and modular, making it simple to assemble, disassemble, and reconfigure them as needs change over time. Stakeholders can streamline modular components for standardization, interchangeability, and compatibility using parametric modeling and simulation, enabling effective assembly and disassembly procedures. When a structure reaches the end of its life, this modular approach not only makes maintenance and refurbishment easier, but it also makes material recovery and reuse easier (Tolio et al., 2017).

Designing for Material Reuse and Deconstruction

Stakeholders can design infrastructure and buildings with deconstruction and material reuse in mind owing to digital twins (Saad et al., 2024). The implementation of techniques like component labeling, reversible connections, and standardized interfaces can help designers recover precious materials and components more quickly during disassembly. With the use of digital twins, it is possible to create digital libraries or catalogs that efficiently record the characteristics and provenance of building components for use in other projects (Çetin et al., 2021).

Resource Allocation and Efficiency

In order to maximize resource utilization efficiency while minimizing waste creation and environmental effects, resource management and optimization are crucial elements of circular building methods. Digital twins function as dynamic platforms in this architecture, coordinating the flow of information, data, and resources throughout the construction lifespan to help stakeholders make well-informed decisions and put circularity and sustainability policies into practice.

Monitoring in Real-Time and Integration of Data

Digital twins integrate many different sources of data, such as sensors, Internet of Things (IoT) devices, and construction equipment, to give stakeholders current time information about how assets are used and performing. Digital twins let stakeholders track and analyze utilization of resources trends, spot ineffectiveness, and make decisions based on data to optimize resource allocation and utilization by gathering data on material movements, energy consumption, and worker productivity (Kineber et al., 2023).

Inventory Control and Material Tracking

Through the digitization of inventory management and material tracking procedures, digital twins enable stakeholders to have visibility and traceability across the whole supply chain. Digital twins allow stakeholders to follow the movement of materials and components from purchase through installation and end-of-life disposal by giving them unique IDs. Effective inventory management, reduced risk of material shortages or overordering, and easier detection of excess resources that can be recycled or put to better use are all made possible by this openness (Jesus et al., 2024).

Enhancement of Building Procedures

By using sophisticated scheduling and simulation techniques, digital twins give stakeholders the ability to optimize construction operations. Digital twins help stakeholders find bottlenecks, expedite processes, and optimize construction schedules to reduce downtime and increase productivity by analyzing data on project timeframes, resource availability, and site conditions. This proactive approach to project management lowers resource waste and the environmental effect of building activities while simultaneously increasing efficiency (Wang et al., 2024).

Transparency in the Supply Chain and Circular Procurement

Stakeholders can learn more about the origin, makeup, and environmental impact of building materials and components by digitizing material passports and product information. With this information at their disposal, procurement specialists can choose materials with high percentages of recycled content, recyclability, and sustainability certifications, increasing the demand for circular processes and goods.

Lifecycle Analysis and Environmental Effects Evaluation

By analyzing the environmental impact of materials and products at every stage of production from extraction and manufacturing to use and disposal, digital twins allow stakeholders to perform thorough lifecycle evaluations. A range of materials and products can have their sustainability credentials compared and evaluated by stakeholders by including environmental performance data into digital twin settings. This allows for the consideration of issues like resource depletion, energy usage, and embodied carbon. Selecting materials with lower environmental footprints and more potential for circularity is made easier with this all-encompassing approach to procurement (Rojek et al., 2020).

Material Passports and Product Information

Within a centralized database, digital twins make it easier to create and manage digital material passports and product information. Material passports enable transparency and traceability throughout the supply chain by giving stakeholders comprehensive information on the make-up, provenance, and lifecycle of building materials and components. Through the digitization of material information, stakeholders may confirm the legitimacy and sustainability credentials of materials, guaranteeing adherence to the criteria and needs of circular procurement (Jensen et al., 2023).

Collaboration with Suppliers and Verification of Certification

Stakeholders can interact with manufacturers, suppliers, and certification agencies at any point in the procurement process by using digital twins as collaborative platforms. Stakeholders can confirm that suppliers and products meet sustainability certification requirements by integrating supplier databases and certification registries into digital twin environments. This ensures that the objectives of circular procurement are met. Adopting sustainable and moral practices is encouraged by this cooperative approach, which builds confidence, accountability, and compliance throughout the supply chain (Tota et al., 2024).

Circular Design Guidelines and Conditions

With the use of digital twins, procurement procedures can more easily incorporate circular design criteria and needs, directing manufacturers and suppliers towards environmentally friendly and circular solutions. Stakeholders can convey their sustainability expectations and goals to suppliers and encourage the adoption of creative and eco-friendly activities by incorporating circularity requirements into procurement contracts and specifications. By empowering stakeholders to generate positive change across the supply chain and encourage the creation and adoption of circular products and materials, this proactive approach to procurement promotes sustainability.

Controlling Waste and Valorization

Digital twins provide a paradigm change from reactive to proactive techniques in the field of waste management. Stakeholders can find opportunities for waste reduction, reuse, and recycling at every stage of the construction process by digitizing waste streams and material inventories (Haris et al., 2024). By identifying excess materials that can be used on-site or diverted to other projects, real-time monitoring and analysis can reduce landfill disposal and maximize the value of underutilized resources.

Digitalization and Tracking of Waste Streams

Through the use of digital twins, stakeholders can track trash generation throughout the construction lifetime and digitize waste streams. Stakeholders can get up-to-date information on garbage types, volumes, and disposal techniques by including sensors, IoT devices, and waste monitoring systems into digital twin settings. With this fine-grained transparency, stakeholders can spot inefficiencies, monitor their progress towards waste reduction goals, and adjust their waste management tactics accordingly.

Sorting and Separation of Materials

Stakeholders can maximize resource recovery and recycling rates by using digital twins to help with the sorting and segregation of waste from construction and demolition projects. Through the digitization of waste streams and material inventories, material properties can be categorized by composition, condition, and recyclability for stakeholders (Khan et al., 2024). Digital twins help with sorting, processing, and disposal decisions by offering insights on the possible worth and reuse potential of certain materials. The utilization of digital twins enables all parties involved in the construction process to apply waste reduction and prevention tactics at each level. Aspects such as material optimization, lean construction techniques, and disassembly-friendly design can be identified by stakeholders through real-time monitoring and analysis. By utilizing digital twins, stakeholders can explore various scenarios and assess the impact of design, procurement, and construction decisions on waste production.

Reusing and Repurposing Materials

The identification and valuation of excess materials that can be recycled or used for other projects is made easier with the use of digital twins. Developing digital catalogs or libraries of construction parts and materials enables stakeholders to inventory surplus resources and determine their suitability for recycling or reuse. Digital twins facilitate the exchange and reuse of materials within a circular economy framework by allowing stakeholders to match leftover materials with pertinent initiatives or stakeholders.

From Linear to Circular Construction Models

A significant change in the planning, construction, and operation of infrastructure and buildings may be seen in the shift from linear to circular construction paradigms. The progression of construction paradigms follows a path from conventional linear models marked by waste production and resource depletion to more sustainable circular approaches driven by lifecycle thinking, resource efficiency, and closed-loop system concepts.

Linear Construction Model

In the past, most methods of construction were linear, depending on the "take-make-dispose" philosophy, which emphasizes the gathering of essential elements, the manufacturing of structure elements, the construction of infrastructure and buildings, and ultimately the demolition and disposal of these structures. The dwindling supply of renewable resources and the buildup of garbage in landfills are caused by this linear model, which is characterized by shortages of resources, decline in the environment, and the production of waste.

Mining and Depletion of Resources

Linear construction models rely heavily on the extraction of natural resources, such as minerals, metals, timber, and fossil fuels, to meet the demands of construction projects. These reserves are frequently gathered without consideration for long-term sustainability, leading to reduction of natural reserves, niche destruction, and ecosystem dislocation. The birth process contributes to environmental declination, including soil corrosion, water pollution, and niche loss, impacting biodiversity and ecosystem services.

Generation and Disposal of Waste

During the course of building, linear construction models are linked to a considerable amount of trash production. Large amounts of garbage are produced during construction and demolition projects. These wastes, which include extra materials, packaging, and debris, are frequently dumped in landfills or incinerators. This method exacerbates the loss of limited landfill space, wastes resources excessively, and pollutes the environment. In addition, there are health and environmental hazards associated with the disposal of building waste, such as the release of greenhouse gasses and contaminated water and soil.

Restricted Lifecycle Theory

Sustainability and resilience over the long term are usually given less weight in linear construction models than immediate cost and schedule concerns. Buildings and infrastructure are frequently not given enough thought throughout their whole lifecycle, from planning and development to use and end-of-life scenarios (Khan et al., 2024). Consequently, there are frequently missed chances to maximize resource efficiency, reduce waste, and improve resilience and adjustability. A building or infrastructure that is not lifecycle-minded may be inefficient, unsustainable in terms of the environment, and more likely to decay and become obsolete too soon. There are major social and environmental consequences associated with the use of linear construction models. Linear construction methods lead to habitat degradation, socioeconomic injustices, biodiversity loss, air and water pollution, and resource depletion and waste generation. Raw material extraction frequently takes place in environmentally delicate places, uprooting indigenous communities and threatening cultural heritage. Furthermore, marginalized groups are disproportionately affected by the disposal of construction waste, which results in environmental injustice and health inequities.

Drivers of Change

The transition to circular building models has been facilitated by a number of factors, including increased public awareness of environmental concerns, legislative demands, market demand for sustainable solutions, and technical developments. A growing number of stakeholders in the construction sector, including contractors, developers, designers, and legislators, are realizing how important it is to implement more sustainable methods that reduce their negative effects on the environment, maximize resource efficiency, and foster long-term resilience. The need for further sustainable construction styles has increased due to growing mindfulness of environmental challenges similar to resource reduction, biodiversity loss, and climate change. The need to limit carbon emigrations, cover the terrain, and save natural coffers for unborn generations is getting increasingly apparent to stakeholders in the structure business. Demand for indirect construction results that put resource effectiveness, waste reduction, and ecosystem preservation first is being driven by this rise in environmental knowledge.

The demand for sustainable buildings and infrastructure is being influenced by shifting consumer tastes and market forces (Waqas Khan et al., 2024). When making selections about what to buy, more and more homeowners, renters, and investors are giving sustainability factors like indoor air quality, energy efficiency, and environmental performance top priority. Developers and investors are motivated to adopt circular construction processes because sustainable structures often lead to higher rents, attract premium tenants, and boost property values helping them stand out in the market. Rapid advancements in technology, including digitalization, robotics, and material science, are unlocking new possibilities for sustainable construction (Saad et al., 2024). Inventions similar to Building Information Modeling (BIM), digital twin, 3D printing, and prefabrication are revolutionizing the way structures are designed, constructed, and operated. These technologies empower stakeholders to optimize resource use, minimize waste, and enhance the sustainability performance of structures and structure throughout their lifecycle. Also, arising accouterments and construction ways, similar as finagled timber, recycled summations, and green concrete (Zaland et al., 2024), offer sustainable druthers to conventional structure accouterments, further driving the relinquishment of indirect construction practices.

Circular Construction Principles

The concepts of circularity, which seek to maximize the lifespan of infrastructure and structures while minimizing waste production and optimizing resource use, serve as the foundation for circular construction models. The use of modular and flexible building methods, the promotion of circular procurement procedures, material selection based on lifecycle performance, and design for disassembly and reuse are important concepts. These guidelines provide an emphasis on resource conservation, promote creativity in building materials and techniques, and encourage cooperation throughout the supply chain (Rasheed et al., 2023).

Lifecycle Performance-Based Material Selection

Indirect construction emphasizes the selection of accouterments grounded on their lifecycle performance, including environmental impact, continuity, recyclability, and renewability. Stakeholders prioritize accouterments with lower environmental vestiges, similar as reclaimed, renewable, and locally sourced accouterments, over conventional druthers. LCA tools and environmental product affirmations (EPDs) are used to estimate the environmental impact of accouterments throughout their lifecycle, from birth and product to use and disposal, enabling informed decision timber and optimization of resource use.

Using Modular and Adaptable Building Techniques

In order to enable infrastructure and structures to change and adapt over time to new conditions and demands, circular construction encourages the use of modular and flexible construction methods. Fast construction, remodeling, and repurposing are made possible by modular components, such as prefabricated panels, modules, and systems, which are made for simple assembly, disassembly, and layout (Lopes et al., 2024). Structure longevity and value are increased, resource efficiency and circularity are encouraged, and there is less need for destruction and disposal when a structure is flexible.

Encouraging the Use of Circular Procurement Methods

Promoting circular procurement strategies that provide preference to materials and goods with high recycled content, recyclability, and environmental certifications is a key component of circular construction. In order to specify sustainable materials and products that satisfy circularity requirements and complement project objectives, stakeholders work in conjunction with manufacturers and suppliers. Throughout the supply chain, material passports and digital documentation offer transparency and traceability, empowering stakeholders to confirm the legitimacy and sustainability credentials of materials and goods and make well-informed purchasing decisions.

Digital Twin Integration for Lifecycle Management

Due to the ability to offer a dynamic platform for lifecycle management, optimization, and innovation, digital twins are essential to circular construction. By combining real-time data, analytics, and simulations, digital twins build virtual versions of physical assets and visualize, evaluate, and optimize each step of the building process from design and procurement through operation and end-of-life. The construction sector is seeing constant innovation and development as a result of this digitalization, which makes it possible for stakeholders to find opportunities for resource optimization, waste reduction, and circularity enhancement (Rasheed et al., 2024).

Benefits and Impacts

There are many advantages to abandoning indirect construction techniques, such as less of an adverse influence on the environment, more efficient use of resources, lower costs, and improved adaptability. Indirect construction models help to save natural resources, lower the export of hothouse gases, and mitigate climate change by minimizing waste output, optimizing material use, and encouraging exercise and recovery (Zaland et al., 2024). Additionally, indirect building methods foster innovation, create new business opportunities, and improve the general sustainability and competitiveness of construction labor.

Material Recovery, Recycling, and Reuse: By encouraging these methods, circular building techniques maximize resource utilization, minimize waste production, and lessen environmental effect.

Carbon Reduction: Building performance can be improved, energy use can be optimized, and greenhouse gas emissions can be decreased with the use of digital twins, all of which help to fight climate change.

Ecosystem Preservation: Through minimizing habitat destruction, biodiversity loss, and ecosystem degradation brought on by resource extraction and waste disposal, circular construction supports ecological sustainability.

Economic Benefits

Cost Savings: Over time, circular building methods save stakeholders money by reducing material waste, operating costs, and lifetime charges.

Revenue Generation: By opening up possibilities for brand-new business models and revenue streams including predictive maintenance, product-as-a-service, and data analytics services, digital twins help stakeholders make more money.

Market Differentiation: Adopting digital twins and circularity makes buildings and infrastructure more marketable and competitive, drawing in high-end tenants, investors, and clients that value sustainability and innovation.

Sustainability Transition: Giving up digital twins and indirect construction methods speeds up the shift to more flexible, regenerative, and sustainable built environments. This greatly helps in the accomplishment of Sustainable Development Goals (SDGs) of the UN Organisation.

Social Benefits

Health and wellbeing: Indirect construction methods for health and well-being improve interior environments, increase occupant comfort, and improve the standard of living for communities and residents.

Fairness and Growth: Digital twins facilitate transparent communication, teamwork, and knowledge exchange among construction workers, providing stakeholders from diverse backgrounds with equal access to information, opportunities, and resources.

Safety and Flexibility: Indirect construction and digital twin improve a structure's adaptability and safety by lowering the risk of natural disasters, climate change, and unforeseen events, as well as by enhancing the safety and well-being of nearby residents and communities.

Technological Benefits

Innovation and Technological Progress: Digital twins encourage the development of novel tools, looks, and outcomes that improve performance, sustainability, and productivity in the building industry.

Data-acquainted Forming Judgements: Well-informed opinions and the optimization of structural processes are facilitated by digital twins, which provide stakeholders with practical perceptivity, predictive analytics, and decision support tools based on real-time data and simulations.

Integration and Interoperability: As digital twins make interoperability and integration among many systems, platforms, and stakeholders easier; they facilitate seamless communication, cooperation, and collaboration throughout the construction lifecycle.

Circular Economy Strategies Enhanced by Digital Twins

The circular economy can be significantly enhanced by the use of digital twins. To achieve the circular economy, it can play certain significant roles in various project phases during the course of a building project's life cycle. Construction can effectively solve obstacles or issues related to the circular economy, minimize waste, and boost salvage value through the integration of digital twins and the circular economy. A few ways in which digital twins could aid in the growth of the circular economy are as follows:

- By optimizing and tracking resources, which would cut down on waste and greenhouse gas emissions while also facilitating effective lifecycle management.
- Extending product life, tracking resources, and facilitating predictive maintenance. This has the potential to reduce material consumption even further, enhance product reusability, and facilitate improved machinery, infrastructure, and consumer products recycling and end-of-life recovery.
- Removing major obstacles related to market inefficiencies that could be impeding the growth or implementation of circular processes, such as, for instance:
  - Inadequate details regarding the circumstances and composition of the product that hinder recycling or reuse.
  - Transaction costs related to locating surplus or secondary material suppliers.
  - The requirement to monitor important parts as they pass through the value chain.

The Significance of Digitalization in Quickening Sustainability

Supply Chain Optimization

Effective management, strategic coordination, and ongoing improvement of interdependent processes are all necessary for supply chain optimization. This journey of optimization seeks to transform business operations by utilizing digital tools and promoting increased transparency throughout the supply chain network. The improvement of demand planning where digital technologies enable organizations to precisely foresee and anticipate market needs is at the heart of this endeavor (Ali et al., 2024). Corporations can maximize stock levels, improve schedules for manufacturing, and guarantee timely shipment of products and services by utilizing advanced analytics and insights based on data. Increased transparency into current procedures is fostered by the inclusion of digital instruments. A comprehensive awareness of every step of the supply chain, from the acquisition of initial supplies to the distribution of finished products, is provided to stakeholders by continuous surveillance and analytics. This increased openness makes it possible to respond quickly to new possibilities or difficulties as well as make proactive decisions. Importantly, the goal of supply chain optimisation goes beyond increasing productivity to include environmental sustainability. Businesses can reduce waste and emissions across the supply chain lifecycle by utilizing digital technologies. To reduce environmental effect while increasing resource efficiency, this calls for implementing eco-friendly behaviors, streamlining transportation routes, and embracing circular economy concepts.

Advanced Process Control

The implementation of advanced process control approaches is a crucial tactic for reducing process unpredictability and energy consumption, both of which contribute to increased operational efficiency (Rasheed et al., 2023). These complex control mechanisms allow systems to operate with little assistance from humans by optimizing processes and laying the foundation for autonomous operations. Through the application of advanced technology and analytical methods, like real-time data analysis and predictive modeling, industries can optimize their processes to function within exact boundaries, resulting in increased efficiency and resource management (Saad et al., 2024). As a result, advanced process control integration promotes sustainability and gives businesses the capacity to precisely and nimbly adjust to changing market conditions.

Predictive and Prescriptive Maintenance

By utilizing machine learning techniques, predictive and prescriptive maintenance can foresee equipment degradation weeks or even months before major problems manifest. These advanced algorithms utilize previous data, sensor readings, and other pertinent parameters to predict possible failures or degradation trends with impressive accuracy. With the help of this proactive strategy, stakeholders can reduce risks and avoid unscheduled downtime by acting in advance. Businesses can save money on operations disruptions and reduce related safety and environmental risks by resolving problems before they get out of hand (Rasheed et al., 2023). Organizations may maximize efficiency, safety, and sustainability throughout their operations by using predictive and prescriptive maintenance to get ahead of equipment breakdowns.

Democratizing Circular Construction Through Digital Twins

The idea of using digital twins to standardize indirect construction is a revolutionary approach to construction diligence. Sustainable construction techniques can be more successfully engaged by stakeholders in all situations through the integration of digital binary technology with indirect frugality principles. The processes by which digital twins can facilitate the democratization of indirect construction are examined in this part, with a focus on efficacy, inclusivity, and transparency.

Circular Economy Principles in Construction

The indirect frugality concept emphasizes minimizing waste, renewing natural systems, and maintaining the usefulness of goods and accessories. In terms of construction diligence, this entails tactics like resource conservation throughout a structure's existence, designing for disassembly, and recycling and reusing accessories. The implementation of indirect frugality principles in the construction industry has the potential to mitigate environmental impacts, optimize resource efficiency, and foster sustainable development.

Synergizing Digital Twins and Circular Economy

Construction practices can be revised by combining digital and indirect thrifty ideas in the following ways:

Better Lifecycle Administration: Digital twins allow for precise shadowing of accessories and elements from design to obliteration. This guarantees that accessories can be properly recycled or reclaimed, facilitating indirect lifecycle operation in compliance with standards like ISO 59010, which emphasizes textiles and indirect frugality concepts as shown in Table 1.

Data-Supported Decision Making: Digital twins provide stakeholders with real-time data and predictive analytics, enabling them to make sustainable decisions throughout every stage of the construction process. This can involve following ISO 59004's recommendations for implementing indirect frugality principles in associations, which include

Table 1: Relationship between Digital Twins, Circular Economy, and ISO standards.

Digital Twins	**ISO 59004**	**ISO 59010**	**ISO 59020**
Definition	Defines what the circular economy is, including its vision, principles, and general guidance including how it can be implemented and contribute to sustainable development.	Provides business-oriented guidance on how to transition towards circular business models and circular value networks by setting goals, identifying circularity aspects to be addressed, and taking actions.	Provides a structured approach to measure and assess circularity performance and sustainability impacts based on standard indicators and complementary methods.
Enhanced Lifecycle Management	**Application**	**Application**	**Application**
Enables detailed tracking of materials and components from design through demolition, ensuring materials can be effectively reused or recycled.	Aligns with the principles and vision of ISO 59004 to ensure sustainable development in construction projects.	Utilizes guidance from ISO 59010 to transition towards circular business models in construction.	Uses ISO 59020 to measure and assess circularity performance and sustainability impacts throughout the construction lifecycle.
Data-Driven Decision Making	**Application**	**Application**	**Application**
Provides real-time data and predictive analytics to empower stakeholders to make sustainable choices throughout the construction process.	Supports the implementation of circular economy principles as outlined in ISO 59004.	Aids in setting goals and identifying circularity aspects as guided by ISO 59010.	Facilitates structured assessment of circularity performance in line with ISO 59020.
Transparency and Collaboration	**Application**	**Application**	**Application**
Facilitates transparent communication among all stakeholders, fostering a collaborative environment to achieve circular economy goals.	Promotes the vision and principles of ISO 59004 by enhancing stakeholder collaboration.	Ensures business-oriented actions are aligned with the guidance of ISO 59010.	Enhances the measurement and assessment process in accordance with ISO 59020.

Improved Design and Construction Processes	**Application**	**Application**	**Application**
Allows for virtual testing and simulation of different design scenarios, enabling the selection of sustainable and efficient options.	Ensures design and construction processes are aligned with the sustainable development goals of ISO 59004.	Utilizes ISO 59010 guidance to transition to innovative, circular construction methods.	Measures and assesses the sustainability impacts of improved design and construction processes in line with ISO 59020.
Maintenance and Deconstruction	**Application**	**Application**	**Application**
Provides insights into the condition and performance of buildings over time, aiding in maintenance and eventual deconstruction for material reuse.	Supports the sustainable maintenance and deconstruction practices advocated by ISO 59004.	Aligns with the business-oriented guidance of ISO 59010 to address circularity aspects.	Uses ISO 59020 to assess the circularity performance of maintenance and deconstruction activities.

minimizing waste, optimizing resource utilization, and choosing eco-friendly accessories.

Collaboration and Transparency: Digital technologies facilitate clear communication between all parties involved, including engineers, masterminds, contractors, and visitors. Table 1 illustrates ISO 59020, which provides guidelines for measuring and evaluating indirect frugality performance. This cooperative environment fosters a shared commitment to indirect frugality pretensions.

Better Design and Construction Processes: By enabling virtual testing and simulation of various design scripts, digital twins make it possible to select the most efficient and sustainable solutions. This may result in cutting-edge building methods that give indirect frugality principles top priority.

Preservation and Demolition: Digital twins provide invaluable insight on the state and functionality of structures over time, facilitating preservation and eventual demolition. This completes the circle in the construction lifetime by guaranteeing that accessories can be recovered and utilized again.

Case Studies

A number of innovative solutions show how digital components will eventually advance indirect construction.

The Edge, Amsterdam: By covering and optimizing energy use, this smart building uses digital binary technology, which results in notable reduction in waste and consumption. Recycled accessories are also used in the architecture of the building, demonstrating the concepts of indirect frugality.

London's Circular Structure: Designed by Arup, this building was created for disassembly and reuse, supported by a digital twin that provides comprehensive documentation of all components and fittings. This method guarantees that the accessories can be reused and the structure may be dismantled fluidly.

Helsinki's Smart City Initiative: Helsinki is creating a smart megacity framework that prioritizes sustainability by utilizing digital twins. Through the incorporation of indirect frugality methods, the megacity hopes to lower its carbon footprint and encourage efficient resource utilization.

References

Ali, Z., Saad, S., Rasheed, K. and Ammad, S. (2024). AI future perspectives and trends in construction. In *AI in Material Science* (pp. 239–261). CRC Press. https://doi.org/10.1201/9781003438489-11.

Bilal, M., Iftikhar Ali, M., Rasheed, K. and Saad, S. (2023). Exploring adoption trends of knowledge management systems in the Software Industry's SMEs: A comprehensive investigation of enablers and barriers. *2023 4th International Conference on Data Analytics for Business and Industry (ICDABI),* 598–604. https://doi.org/10.1109/ICDABI60145.2023.10629373.

Birat, J. -P. (2015). Life-cycle assessment, resource efficiency and recycling. *Metallurgical Research & Technology,* 112(2): 206. https://doi.org/10.1051/metal/2015009.

Boje, C., Hahn Menacho, Á. J., Marvuglia, A., Benetto, E., Kubicki, S. et al. (2023). A framework using BIM and digital twins in facilitating LCSA for buildings. *Journal of Building Engineering,* 76: 107232. https://doi.org/10.1016/j.jobe.2023.107232.

Çetin, S. (2023). Towards a circular building industry through digitalisation. *Architecture and the Built Environment.* https://doi.org/10.59490/ABE.2023.22.7284.

Çetin, S., De Wolf, C. and Bocken, N. (2021). Circular digital built environment: An emerging framework. *Sustainability,* 13(11): 6348. https://doi.org/10.3390/su13116348.

Chauhan, C., Parida, V. and Dhir, A. (2022). Linking circular economy and digitalisation technologies: A systematic literature review of past achievements and future promises. *Technological Forecasting and Social Change,* 177: 121508. https://doi.org/10.1016/j.techfore.2022.121508.

de Wit, M., Haigh, L. and Von Daniels, C. (2020). *Closing the Circularity Gap in the Netherlands Circularity Gap Report.* https://www.circularity-gap.world/netherlands.

den Elzen, M. G. J., Dafnomilis, I., Forsell, N., Fragkos, P., Fragkiadakis, K. et al. (2022). Updated nationally determined contributions collectively raise ambition levels but need strengthening further to keep Paris goals within reach. *Mitigation and Adaptation Strategies for Global Change,* 27(5): 33. https://doi.org/10.1007/s11027-022-10008-7.

Eladly, A. M., Abed, A. M., Aly, M. H. and Salama, W. M. (2023). Enhancing circular economy via detecting and recycling 2D nested sheet waste using Bayesian optimization technique based-smart digital twin. *Results in Engineering,* 20: 101544. https://doi.org/10.1016/j.rineng.2023.101544.

Fallah Shayan, N., Mohabbati-Kalejahi, N., Alavi, S. and Zahed, M. A. (2022). Sustainable Development Goals (SDGs) as a framework for Corporate Social Responsibility (CSR). *Sustainability,* 14(3): 1222. https://doi.org/10.3390/su14031222.

Ford, G., Bartley, T., Igba, J., Turner, A., McMahon, C. et al. (2013). *Product Life Cycle Data Management: A Cross-Sectoral Review* (pp. 58–67). https://doi.org/10.1007/978-3-642-41501-2_7.

Haris, M., Saad, S., Ammad, S. and Rasheed, K. (2024). AI in fabrication and construction. In *AI in Material Science* (pp. 169–192). CRC Press. https://doi.org/10.1201/9781003438489-8.

Haris, M., Saad, S., Ammad, S., Rasheed, K., Aman, H. et al. (2024). Integrating sustainable practices for effective solid waste management in peshawar: Overcoming challenges and promoting environmental. *2024 ASU International Conference in Emerging Technologies for Sustainability and Intelligent Systems (ICETSIS),* 1–7. https://doi.org/10.1109/ICETSIS61505.2024.10459445.

Hossain, M. U., Ng, S. T., Antwi-Afari, P. and Amor, B. (2020). Circular economy and the construction industry: Existing trends, challenges and prospective framework for sustainable construction. *Renewable and Sustainable Energy Reviews,* 130: 109948. https://doi.org/https://doi.org/10.1016/j.rser.2020.109948.

Jensen, S. F., Kristensen, J. H., Adamsen, S., Christensen, A., Waehrens, B. V. et al. (2023). Digital product passports for a circular economy: Data needs for product life cycle decision-making. *Sustainable Production and Consumption*, 37: 242–255. https://doi.org/10.1016/j.spc.2023.02.021.

Jesus, V., Kalaitzi, D., Batista, L. and Lopez, N. (2024). *Digital Twins of Supply Chains: A Systems Approach*. https://doi.org/10.36227/techrxiv.24073527.v2.

Joshi, A. D. and Gupta, S. M. (2019). Evaluation of design alternatives of End-of-Life products using Internet of Things. *International Journal of Production Economics*, 208: 281–293. https://doi.org/10.1016/j.ijpe.2018.12.010.

Khan, M. I., Khan, N., Saad, S. and Khan, M. W. (2024). Artificial Intelligence in construction materials. In *AI in Material Science* (pp. 47–64). CRC Press. https://doi.org/10.1201/9781003438489-3.

Kineber, A. F., Singh, A. K., Fazeli, A., Mohandes, S. R., Cheung, C. et al. (2023). Modelling the relationship between digital twins implementation barriers and sustainability pillars: Insights from building and construction sector. *Sustainable Cities and Society*, 99: 104930. https://doi.org/10.1016/j.scs.2023.104930.

Lewis, H., Gertsakis, J., Grant, T., Morelli, N., Sweatman, A. et al. (2017). *design + environment*. Routledge. https://doi.org/10.4324/9781351282208.

Lopes, L., Penazzato, L., Reis, D. C., Almeida, M., Oliveira, D. V. et al. (2024). A holistic modular solution for energy and seismic renovation of buildings based on 3d-printed thermoplastic materials. *Sustainability*, 16(5): 2166. https://doi.org/10.3390/su16052166.

Meng, X., Das, S. and Meng, J. (2023). Integration of Digital Twin and circular economy in the construction industry. *Sustainability*, 15(17). https://doi.org/10.3390/su151713186.

Milios, L. (2021). Overarching policy framework for product life extension in a circular economy—A bottom-up business perspective. *Environmental Policy and Governance*, 31(4), 330–346. https://doi.org/10.1002/eet.1927.

Moghrabi, I. A. R., Bhat, S. A., Szczuko, P., AlKhaled, R. A., Dar, M. A. et al. (2023). Digital transformation and its influence on sustainable manufacturing and business practices. *Sustainability*, 15(4): 3010. https://doi.org/10.3390/su15043010.

Mügge, J., Grosse Erdmann, J., Riedelsheimer, T., Manoury, M. M., Smolka, S. -O. et al. (2023). Empowering end-of-life vehicle decision making with cross-company data exchange and data sovereignty via Catena-X. *Sustainability*, 15(9): 7187. https://doi.org/10.3390/su15097187.

Mügge, J., Seegrün, A., Hoyer, T. -K., Riedelsheimer, T., Lindow, K. et al. (2024). Digital Twins within the circular economy: Literature review and concept presentation. *Sustainability*, 16(7): 2748. https://doi.org/10.3390/su16072748.

Numanovich, A. I. and Abbosxonovich, M. A. (2020). Utilising natural resources has always been done in a linear fashion, with the idea being that resources are acquired, used, and then disposed of as trash. The sustainability of the natural environment was seriously challenged by this "take-make-dispose". *EPRA International Journal of Multidisciplinary Research (IJMR)-Peer Reviewed Journal*, 2: 198–210. https://doi.org/10.36713/epra2013.

Oliveira, J., Schreiber, D. and Jahno, V. (2024). *Circular Economy and Buildings as Material Banks in Mitigation of Environmental Impacts from Construction and Demolition Waste*. https://doi.org/10.20944/preprints202405.1353.v1.

Patwa, N., Sivarajah, U., Seetharaman, A., Sarkar, S., Maiti, K. et al. (2021). Towards a circular economy: An emerging economies context. *Journal of Business Research*, 122: 725–735. https://doi.org/10.1016/j.jbusres.2020.05.015.

Rasheed, K., Saad, S. and Ammad, S. (2023). Green financing for sustainable economic growth: exploring critical factors towards resilient economies. *2023 International Conference on Sustainable Islamic Business and Finance (SIBF)*, 104–111. https://doi.org/10.1109/SIBF60067.2023.10379984.

Rasheed, K., Saad, S., Ammad, S. and Badshah, I. (2023). IoT enhanced solar device management: Integrating remote surveillance and energy efficiency for sustainable operation. *2023 TRON Symposium (TRONSHOW)*, 1–6.

Rasheed, K., Saad, S., Ammad, S. and Bilal, M. (2023). Integrating sustainability management and lean practices for enhanced supply chain performance: Exploring the role of process optimization in SMEs. *Engineering Proceedings*, 56(1). https://doi.org/10.3390/ASEC2023-16370.

Rasheed, K., Saad, S., Zaland, A., Khan, M. W., Ali, Z., Haris, M., Ammad, S. and Sadiq, T. (2024). Circular economy and critical barriers: Mapping the pathways and success metrics for sustainable circular success in industrialised South Asian developing nations. *Results in Engineering*, 24: 102995. https://doi.org/10.1016/j.rineng.2024.102995.

Rasheed, K., Waqas, M., Saad, S., Zaland, A., Ali, Z. et al. (2024). Ethical and societal implications of AI in construction. In *AI in Material Science* (pp. 216–238). CRC Press. https://doi.org/10.1201/9781003438489-10.

Rojek, I., Mikołajewski, D. and Dostatni, E. (2020). Digital Twins in product lifecycle for sustainability in manufacturing and maintenance. *Applied Sciences*, 11(1): 31. https://doi.org/10.3390/app11010031.

Saad, S., Ammad, S. and Rasheed, K. (2024). *AI in Material Science*. CRC Press. https://doi.org/10.1201/9781003438489.

Saad, S., Haris, M., Ammad, S. and Rasheed, K. (2024). AI-assisted building design. In *AI in Material Science* (pp. 143–168). CRC Press. https://doi.org/10.1201/9781003438489-7.

Saad, S., Rasheed, K., Ammad, S., Ali, Z., Zaland, A. et al. (2024). The conclusive role of AI in Industry 4.0 construction materials. In *AI in Material Science* (pp. 262–270). CRC Press. https://doi.org/10.1201/9781003438489-12.

Saad, S., Rasheed, K., Ammad, S., Waqas Khan, M., Zaland, A. et al. (2024). Is AI the architect of tomorrow's materials in the age of Industry 4.0? In *AI in Material Science* (pp. 1–14). CRC Press. https://doi.org/10.1201/9781003438489-1.

Saad, S., Zaland, A., Ammad, S. and Rasheed, K. (2024). Robotics in construction. In *AI in Material Science* (pp. 115–142). CRC Press. https://doi.org/10.1201/9781003438489-6.

Tolio, T., Bernard, A., Colledani, M., Kara, S., Seliger, G. et al. (2017). Design, management and control of demanufacturing and remanufacturing systems. *CIRP Annals*, 66(2): 585–609. https://doi.org/10.1016/j.cirp.2017.05.001.

Tota, R., Facchini, F., Iavagnilio, R. P., Mossa, G., Sassanelli, C. et al. (2024). Digital Twin supporting environmental performance evaluation according to ISO certification: An application case in the tire industry. *Production & Manufacturing Research*, 12(1). https://doi.org/10.1080/21693277.2024.2351839.

Vezzoli, C. A. (2018). *Design for Environmental Sustainability*. Springer London. https://doi.org/10.1007/978-1-4471-7364-9.

Wang, W., Zaheer, Q., Qiu, S., Wang, W., Ai, C. et al. (2024). Digital Twins technologies. In *Digital Twin Technologies in Transportation Infrastructure Management* (pp. 27–74). Springer Nature Singapore. https://doi.org/10.1007/978-981-99-5804-7_2.

Wang, X. V. and Wang, L. (2019). Digital twin-based WEEE recycling, recovery and remanufacturing in the background of Industry 4.0. *International Journal of Production Research*, 57(12): 3892–3902. https://doi.org/10.1080/00207543.2018.1497819.

Waqas Khan, M., Saad, S., Ammad, S., Rasheed, K., Jamal, Q. et al. (2024). Smart infrastructure and AI. In *AI in Material Science* (pp. 193–215). CRC Press. https://doi.org/10.1201/9781003438489-9.

Zaland, A., Saad, S., Ahmad, W., Ammad, S., Khan, M. W. et al. (2024). Enhancing mechanical properties of concrete by adding polypropylene fibers. *2024 ASU International Conference in Emerging Technologies for Sustainability and Intelligent Systems (ICETSIS)*, 771–778. https://doi.org/10.1109/ICETSIS61505.2024.10459663.

Zaland, A., Saad, S., Rasheed, K. and Ammad, S. (2024). Automation in construction industry. In *AI in Material Science* (pp. 88–114). CRC Press. https://doi.org/10.1201/9781003438489-5.

Chapter 7

Integrating Digital Twins and Robotics

Ahmad Zaland,[1] *Zawar Ali,*[1] *Muhammad Haris,*[1]
Syed Saad[2,*] and *Kumeel Rasheed*[1]

Introduction

With the introduction of Industry 4.0, the technological paradigm of human civilization has changed. A wide range of new synergistic fusions of current and newer technologies have been noted to have witnessed a stunning rise in this new era of industry, automation, and man-machine interaction (Nardo et al., 2020). In addition, it has sparked the creation of certain novel ideas like Operator 4.0 and Space Factory 4.0, which are flourishing technological hybrids. In addition to the industrial side, these technologies are also proving to be quite beneficial for the research sectors (Hirsch-Kreinsen, 2016).

Ironically, the term "space robotics" was not even coined when the first digital twin techniques were applied in the 1960s. NASA later termed it "living models". Moreover, space robotics provided the first notable example of DT implementation, which was noted in 1970. It was employed on the illustrious Apollo 13 mission, which featured advanced robotic technologies at the time for crucial operations like docking. Along with being DT's first space robotics success story, the mission saw the

[1] Department of Civil Engineering, CECOS University of IT and Emerging Sciences, Pakistan.
[2] Science, Engineering & Manufacturing Centre Sandwell College West Bromwich, United Kingdom.
* Corresponding author: syed.saad@sandwell.ac.uk

crew safely return from an imminent danger brought on by the oxygen tank explosion aboard the craft thanks to the technology's successful deployment. But unlike the real-time integration of virtual and physical replicas seen today, the DT used back then was a multi-simulator cyber-physical system that primarily concentrated on physical infrastructure. David Gelernter proposed a similar idea in 1991 called "Mirror Worlds," in which computer models use data from the real world to simulate reality (Mazumder et al., 2023).

In 2002, John Vickers of NASA first used the term "digital twin." Vickers and Dr. Michael Grieves worked together at the same time to modify the idea of digital twins to enhance product lifecycle management (PLM) in the manufacturing industry. He called it the "Conceptual Ideal for PLM" at first. Even at this preliminary stage, he listed several crucial characteristics of digital twins (Grieves, 2023). In his article, Grieves examined the differences between actual and virtual settings, highlighting the necessity of data and information sharing between real and virtual entities for them to reflect one another. His "Mirrored Spaces Model" had been replaced with the "Information Mirroring Model" by 2006 (Jones et al., 2020).

The idea of the digital twin had seen a sharp increase in popularity since 2003. The 2020's top strategic technological trend is identified as hyper-automation, with digital twins playing a major part (Wei et al., 2022). This increase in interest is being driven by initiatives like Industry 4.0 and digital futures (Rasheed et al., 2024). Furthermore, a number of significant technological developments, including real-time sensors, big data, and the Internet of Things, have reduced prices. Surveys indicate that about 75% of companies working on IoT initiatives are already deploying digital twins (Javaid et al., 2022). The concept is obviously catching on with early users. Markets and Markets estimated that the value of the digital twin's marketplace in 2023 was USD 10.1 billion and poised to reach USD 110.1 billion by 2028, growing at a compound annual growth rate of 61.3% from 2023 to 2028 (Markets and Markets, 2024).

Holistic Integration Strategies

Businesses in a wide range of sectors have discovered the concept of "digital twins" to be an extremely helpful instrument. It is vital to understand that effective implementation requires much more than just using the technology, even though the technology enabling digital twins is undoubtedly innovative. In order for businesses to fully utilize the potential of digital twins, they need to take a comprehensive approach that includes strategy, design, organizational effect, and a vision for future expansion and innovation (Sharma et al., 2022) as shown in Figure 1.

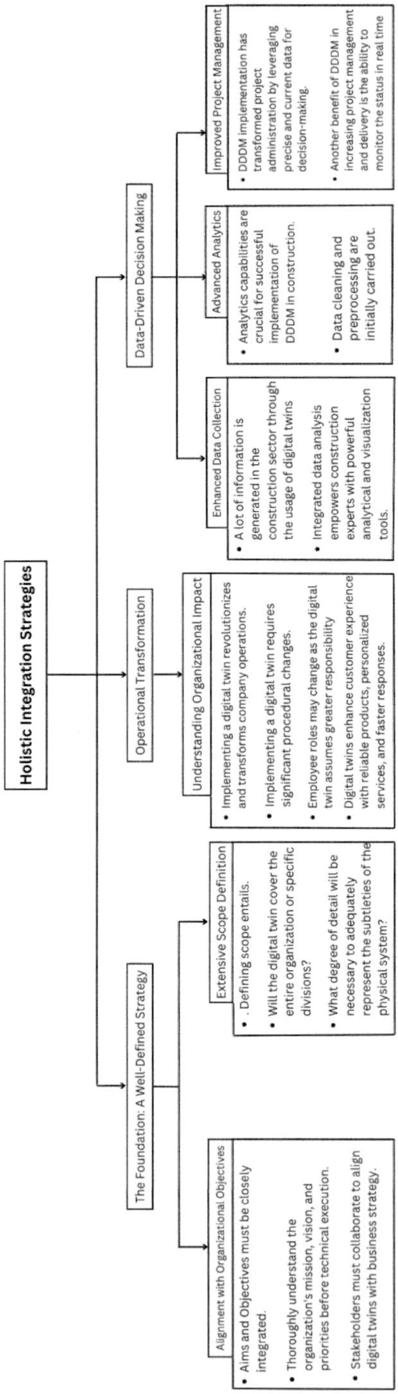

Figure 1: Digital Twin integration strategies.

The Foundation: A Well-Defined Strategy

Starting the process of putting a digital twin into practice requires a comprehensive and well-defined plan. The foundation for maximizing the potential of this cutting-edge technology is laid by this calculated approach. We examine the essential elements of developing a thorough plan for implementing digital twins in this in-depth analysis, stressing the vital significance of alignment with organizational goals, identifying particular difficulties, and carefully defining the scope (Mihai et al., 2022).

Alignment with Organizational Objectives

An organization's overall aims and objectives must be closely integrated with a successful digital twin project. Comprehending the organization's mission, vision, and strategic priorities thoroughly is crucial prior to diving into the technical details of execution. Through coordinating the digital twin project with these overarching goals, interested parties can guarantee that each endeavor makes a significant contribution to the prosperity of the company (Sarker, 2023). Stakeholders need to work together to find ways that digital twin technology can complement and improve current business strategy in order to achieve alignment. This could entail looking into possible areas for development, such as increasing customer happiness, speeding up the time to market for new items, or optimizing operational efficiency. Through the creation of explicit links between the digital twin project and strategic objectives, stakeholders can gain support and buy-in from all areas of the company (Cardoso et al., 2022).

Extensive Scope Definition

Defining the project's scope is the next stage in creating a thorough digital twin strategy after the general goals and particular difficulties have been determined. The scope outlines the parameters of the digital twin project, making it clear which systems, assets, and procedures will be included as well as how they will interact with one another in the virtual setting. Defining scope entails making strategic choices on the extent and complexity of the digital twin deployment. Will the organization as a whole be included in the digital twin, or will it concentrate on particular divisions, locations, or procedures? What degree of detail will be necessary to adequately represent the subtleties of the physical system? It is important to give considerable thought to these issues in order to guarantee that the scope is meaningful and practical (Perno and Hvam, 2020).

Understanding Organizational Impact

Establishing a digital twin in a company is a revolutionary endeavor that can significantly alter many aspects of operations, not just a technology update. Gaining the most out of the digital twin implementation and guaranteeing a seamless transition depend on the ability to comprehend and manage these effects. Adopting new technologies and making considerable modifications to current procedures are required when implementing a digital twin. These adjustments may have a broad impact on anything from long-term strategic planning to daily operations. To incorporate real-time data analytics and predictive maintenance capabilities offered by the digital twin, for example, operational workflows may need to be redesigned. Workers used to conventional procedures would have to pick up new tools, understand intricate data visualizations, and apply insights to decision-making.

Employee roles may change as the digital twin assumes greater responsibility. By leveraging digital twin technology, labor -intensive and manual processes can be automated or enhanced, enabling employees to focus on more valuable tasks (Haris et al., 2024). To avoid problems before they arise, maintenance personnel could, for instance, switch from doing standard inspections to reviewing predictive maintenance reports produced by the digital twin. A more competent and motivated staff can result from this role evolution, but it also calls for cautious change management. Employers should involve staff members in the transition process, offer possibilities for career growth, and convey the changes to them properly. Employees can find it easier to see the benefits of their new positions and responsibilities if a culture of innovation and constant learning is promoted (Javaid et al., 2023).

The customers' experience can be greatly improved by a digital twin by offering more dependable products, individualized services, and quicker response times. It is vital that these advantages are conveyed successfully to clients, nonetheless. It must comprehend the benefits of the digital twin and how it will enhance their relationships with your company. Developing a well-defined communication plan is essential. In-depth descriptions of the digital twin's functionalities, case studies showing effective deployments, and customer endorsements should all be a part of this approach. Furthermore, providing clients with first-hand knowledge of the benefits through demonstrations or trial programmes helps build trust and enthusiasm.

The influence of implementing a digital twin also extends to other stakeholders and the supply chain partners. These other parties will have to communicate with the digital twin, which can change how it interacts with the company. A digital twin can provide supply chain partners with increased visibility and chances for cooperation. Real-time data exchange can shorten lead times, increase inventory control, and optimize the efficiency of the supply chain (Saad et al., 2024). But to do this, firms need a solid structure for sharing data and solid alliances built on mutual respect and benefit.

Data-Driven Decision Making

Data-Driven Decision Making (DDDM) is a managerial responsibility that aims to make conclusive judgments by focusing on actual information and data gathered from an organization. This style places a strong emphasis on using factual information to make business decisions, rather than relying solely on messages. A cyclic process illustrating key stages of DDDM, starting with enhanced data collection and progressing through advanced analytics and real-time monitoring is shown in Figure 2. The application domains of DDDM are extensive and significant for numerous sectors and their actual implementation. The proper utilization of data enhances the overall understanding of organizational operations, customer behavior, and market dynamics. It also facilitates the evaluation of the extent to which goals and objectives have been accomplished. Having this knowledge allows individuals to make well-informed decisions based on statistics, which can improve effectiveness, expand a company's customer base, and ultimately raise the organization's profitability (Nessari et al., 2024).

Currently, the utilization of DDDM has become essential in the present company landscape, regardless of the sector. Businesses are recognizing the benefits of employing analytical tools in their operations to gain a competitive edge. In the retailing segment, for example, organizations can apply predictive analytics for concerns including product offerings, marketing communications, and inventory and distribution. Similarly, in the field of healthcare, big data plays a leading role in shaping the advancement of workflows, organizational structures, treatment techniques, and the evaluation of patient conditions.

An exemplary illustration of the impacts of DDDM can be observed in a practical scenario involving a prominent e-commerce enterprise. The company was compelled to improve its search functionality to strengthen its capacity to provide clients with relevant search results. This platform was also capable of assessing the search strings, customers' decisions, and their browsing behaviors with the use of data analytics. Based on these findings, the management implemented a highly advanced search mechanism that boosted the precision and pertinence of the search

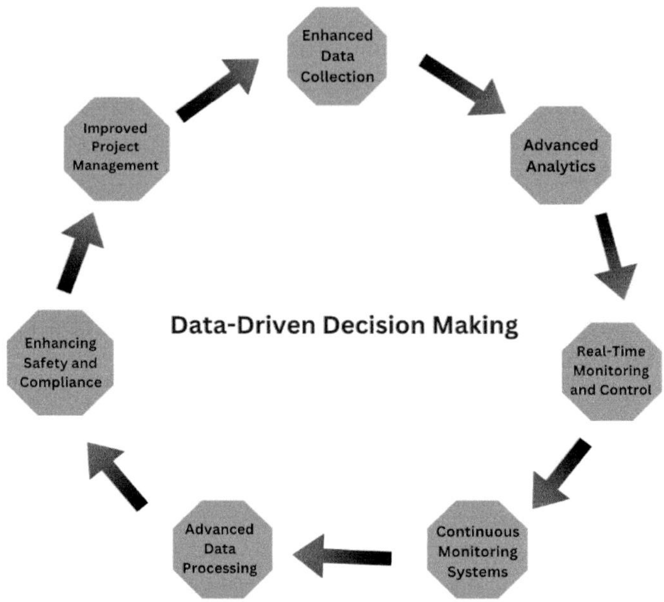

Figure 2: A cyclic process illustrating key stages of data-driven decision making.

outputs. It acted as a bridge which resulted in an increased level of clients contemplating and finally buying the brands, and a better level of happiness (Chen and Bai, 2023).

Enhanced Data Collection

A lot of information is generated in the construction sector through the usage of digital twins, but the information collected is of no use if there is no provision to evaluate the data. In the context of DDDM, integrated data analysis is a significant step due to its potential to allow construction experts to work with analytical and visualization tools. These are for example the prediction of equipment failures and preventive maintenance; the identification of wasteful and inefficient techniques to facilitate their substitution; assessment of construction-related risks through real-time prediction of environmental, worker, or structural health; and planning and scheduling of construction activities with the aid of dependencies and their effects on the project time. In construction, the role of DDDM is to enhance collaborative decision-making among the project managers, engineers, architects, contractor subcontractor clients, and owners via the availability of data to develop confidence and better coordination and decision-making throughout the construction process (Demirdöğen et al.,

2021). However, DDDM is a never-ending process such that construction businesses can also look for possibilities on one project, try to compare it to others, adjust circumstances, and develop a culture where employees from the junior to senior level seek the best ways of enhancing the process (Chen and Bai, 2023).

Advanced Analytics

Analytics' capabilities for DDDM in construction is also a major success factor in executing this framework in the construction industry. Once the extensive data from digital twins across the building site is gathered completely, it undergoes preprocessing and analysis using sophisticated analytics tools and methodologies. To begin the above procedure, crucial steps that include data cleaning and preprocessing are initially carried out. Data that is generated from sources like sensors, IoT devices, as well as the construction equipment that is employed is generally raw, vast, and complex. Data cleaning is the process of eliminating noisy information, fixing errors, and putting the data into an analysis-friendly shape. Building professionals can then use the structured data to perform descriptive analytics, aiming to identify historical trends within the industry (Presciuttini and Portioli-Staudacher, 2024). Data mining results are used primarily to present an overview of events and situations to stakeholders, combined with the application of more advanced methods of study. Hence, whereas descriptive analysis gives valuable generalizations, the following act is performed by the prospective analysis. Therefore, utilizing machine learning (ML) algorithms and AI models in construction organizations can lead to the identification of specific patterns that will assist in predicting the future (Haris et al., 2024). One of the major areas of value revealed in the application of predictive analytics is on the topic of predictive maintenance where it is possible to identify probable breakdowns in various equipments and carry out the maintenance ahead of the damage occurring thereby cutting on a lot of expenditures (Bhattacharyya, 2024).

Improved Project Management

The implementation of DDDM has significantly altered the administration of building projects and has allowed construction specialists to make their decisions based on a lot of precise and current data. Another benefit of DDDM in augmenting project management and delivery is the ability to monitor the status in real time (Rane and Narvel, 2022). Project managers, using data generated from digital twins, will have the tools to monitor construction progress in real-time and compare it to the planned schedule.

Such transparency in ongoing activities enables the early detection of slow-moving areas or planning delays, allowing them to be addressed before causing significant setbacks. Furthermore, DDDM is crucial for the efficient provision of resources and helps to make appropriate judgments in this regard. Utilizing data on usage trends, consumption patterns, and future demand for labor, materials, and equipment, project managers can develop enhanced plans that align with the constraints of the project management triangle (Rasheed et al., 2022). These insights can ensure the availability of the desired resource and minimize extra idle time improving effectiveness.

Case Studies and Practical Applications

Smart Building Projects

In the context of smart building projects, the integration of digital twins has brought a major change in monitoring and managing building systems. Microsoft's smart building initiative at its Puget Sound headquarters, powered by Azure Digital Twins, exemplifies the future of intelligent workspaces (Bay, 2023). Launched in 2021 as part of a comprehensive campus modernization project, this initiative integrates IoT devices and digital twins to create real-time digital replicas of physical environments. These smart buildings autonomously manage energy distribution, guide visitors, and optimize conference room air quality. The platform combines data from motion and occupancy sensors, HVAC systems, and security devices to enhance both employee productivity and facilities management (Bay, 2023). This novel approach aligns software development with construction timelines, ensuring seamless integration and operational efficiency, and is set to expand across Microsoft's global properties, showcasing a pioneering model for sustainable and efficient building management.

Infrastructure Projects

The implementation of the digital twin concept in civil engineering is mostly related to large-scale infrastructures like bridges or tunnels for which constant monitoring of the overall structural health is inevitable. In Los Angeles, the Sofi sports stadium has its own digital twin. It models the stadium and the 300-acre Hollywood Park campus around it. The twin collects data in real-time from each area of the park's operations so that event organizers can improve the customer experience and maintenance (Phillips, 2020). These users engage with the twin using apps that relate to specific features and functions. The use of digital twin technology will make it easier for facilities managers to enhance game days for patrons.

Residential Developments

Skanska, a multinational construction and development company renowned for its extensive global presence, utilizes digital twins to revolutionize its approach to large-scale projects, including residential developments. A digital twin is a virtual representation of a physical asset or system, allowing Skanska to simulate and optimize construction processes in a virtual environment before executing them on-site (De Valence, 2022). This technology enables Skanska to enhance project management by accurately predicting outcomes, optimizing resource allocation, and identifying potential risks early in the construction phase. By integrating digital twins into their workflow, Skanska enhances collaboration among project teams, stakeholders, and clients, ensuring efficient communication and informed decision-making throughout the project lifecycle. This innovative approach not only improves construction efficiency and quality but also reinforces Skanska's commitment to sustainable practices and delivering high-value, customer-centric residential developments globally.

Real-Time Monitoring and Control

Real-time monitoring and control are crucial sub-processes in construction management as it provides an efficient method for construction teams to observe and oversee the dynamic situations within a construction site. This strategy combines the utilization of leading-edge sensors, IoT devices, and synchronized data analysis systems to guarantee optimal monitoring and control of various components of construction (Khan et al., 2024). The implementation of real-time monitoring and control relies on the utilization of construction site models, which are tangible duplicates of the actual construction sites. These models serve as a means to consolidate data for analysis. These digital twins consist of various sensors, including environmental sensors, structural health sensors, equipment tracking sensors, and safety sensors for workers. These sensors are loaded with a significant amount of data and continuously provide input to the digital ecosystem (De Man et al., 2023).

By integrating and analyzing real-time information, construction personnel can now gain an unprecedented amount of visibility into the current status of the project. Temperature, humidity, and rainfall are environmental elements that have an impact on building activity. These parameters can be assessed and adjusted to improve construction efficiency. Structural health sensors can detect indications of stress or deformation, allowing for prompt repairs and the prevention of potential risks. Movable assets can be monitored and tracked to analyze usage trends for predictive maintenance if needed (Waqas Khan et al., 2024).

The potential for implementing real-time monitoring and control in the construction industry is vast. The precise time information is highly valuable in the field of project management as it allows for the identification of potential difficulties that can pose significant challenges shortly. This approach facilitates the management of delays, bottlenecks, and resource shortages within the project. It also enables the implementation of corrective measures to ensure the project cycle remains on track and project expenses are controlled (Radich et al., 2023).

Continuous Monitoring Systems

In the construction process, monitoring systems form the core of the DDDM approach, in which the function of a continuous monitoring system cannot be downplayed. These are essential elements of digital twins and consist of many elements of the construction site measurement and monitoring systems using sensors, IoT devices, and communication interfaces for the collection of real-time information about the construction site. Smart skins embedded in the building structure continuously monitor critical factors like structural stress, strain levels, and temperature. This real-time data allows for the immediate detection and identification of potential issues within the building (Tang et al., 2024).

Environmental monitoring tracks real-time weather conditions, air quality, and environmental data, aiding construction teams in risk management and compliance with project requirements. Smart IoT devices are connected systems that can monitor and send data on the state and activity of equipment and mechanical assemblies, which enhances efficiency and decreases idle time (Rasheed et al., 2024). Smart watches and safety sensors also monitor the movements of the workers and the conditions within the working environment to give feedback alerts that ensure that the working environment is safe. With the help of these continuous monitoring systems, a vast array of real-time data becomes available for construction organizations, which can help the latter make business and operational decisions with improved efficiency, safety, and ever-better control over the construction project from the beginning to the end (Oliva et al., 2024).

Advanced Data Processing

The implementation of monitoring systems in construction projects provides a tremendous volume of data, and adequate data analysis techniques are needed to identify critical performance indicators for subsequent decision-making. One of the key initiatives for improving data processing in the construction business is the usage of edge computing. The above analysis demonstrates that processing data at the edges of construction networks, closer to the data sources, reduces latency and enhances response times, enabling construction organizations to implement prompt intervention. It also provides real-time decision-making since edge processing is handled at the site and substation level, thereby eliminating the requirement for cloud computing to detect anomalies or even start corrective actions. For instance, edge devices attached to mining equipment can analyze data in real-time, and subsequently offer the equipment operators with real-time warnings on defects making it easier to contain damages (Hirano et al., 2024).

Although there is a trend for edge computing to address the aspect of closeness, cloud computing remains vital in handling the analytic and longer-term storage necessary in building projects. Cloud systems provide the scalability required for complex analytical models, artificial intelligence (AI), and extensive data storage (Saad et al., 2024). By collecting data from multiple construction sites, organizations can gain valuable insights into their operations. This includes the efficient use of resources, the time required to complete various projects, and overall performance at an enterprise level (Bilal et al., 2023). The fundamental aspects allowing the latest developments in data processing are AI and ML. The algorithms employ real-time data that is generated by monitoring systems and utilize it to discover trends, foresee implications, and then give algorithm-driven solutions to the working construction teams. For instance, by combining descriptive datasets and recent sensor measurements, the deployment of ML models can anticipate equipment failures and help arrange timely repair to avoid costly equipment breakdowns (Zhang et al., 2024).

Real-Time Control Systems

Real-time control systems represent the pinnacle of continuous monitoring and advanced data processing in construction organizations, enabling them to manage and regulate activities in accordance with site conditions instantaneously. An essential aspect of the management system in real-time control systems is the incorporation of automated control systems in construction sites that use robotics and different automated equipment in construction activity with the help of real-time input from the systems (Zaland et al., 2024). For instance, self-driven construction machines can adjust routes and charge activities according to environmental circumstances, and achieve the highest efficiency and safety if required. An autonomous-working bulldozer may modify its direction and speed depending on the information received from sensors, which boosts the efficiency of operations, lifting and moving terrain elements, and avoiding interference from other equipment, and objects (El-khawaldeh et al., 2024).

A known element of real-time control systems also concerns the usage of maintenance models designed for prevention. This could enable teams to move where they are observing a tool or equipment being used and what its current status is, to not only rely on a typical routine for the maintenance but also be able to proactively take appropriate action. This predictive maintenance model assists in minimizing actual downtime time, avoiding equipment failure that occurs without prior indication, and in general minimizes the overall maintenance cost. A construction site can require multiple cars for construction purposes and these vehicles can be checked periodically for damage symptoms; this way, maintenance can be scheduled only when necessary to maintain the vehicles for longer and improve the effectiveness of operations. The integration of real-time control systems also incorporates flexibility in construction workflows. Real-time data enables construction managers to handle unexpected occurrences on the construction site such as the weather by making changes to the planned construction process. In the scenario where storms are expected, construction managers can modify jobs that would need extra unprotected outdoor labor such as the pouring of concrete, and redirecting personnel to conduct more interior work to maintain project progress even with interruptions (Pei et al., 2024).

Enhancing Safety and Compliance

The implementation of real-time monitoring and control in construction projects has over time kind of increased the safety and compliance approaches of the sector. Such applications are vital in the detection of risks, training on compliance standards, and especially in emergency management. Another advantage of real-time monitoring is the possibility of revealing several risks. Using environmental and worker monitoring, construction sites can identify risks and hazards in time with the help of sensors (Saad et al., 2024). Alerts can subsequently be dispatched to workers and supervisors in the area, prompting them to address and rectify hazardous situations immediately. For example, gas concentration detectors can detect gas leaks in real-time, triggering alarms and prompting the immediate evacuation of workers (Cheng et al., 2022). Through real-time monitoring of the construction operations, these technologies guarantee that all the activities are safe and other legal issues such as environmental conservation are addressed. Compliance issues can be spotted instantaneously, enabling construction teams to rectify the issues on the same day they are discovered. For example, environmental sensors can detect emissions from construction equipment; they stay to regulate the levels of emission to meet environmental regulations and take swift action when the emissions are high. Emergency real-time monitoring solutions are crucial during emergencies, especially when evacuation procedures are underway.

Robotic Autonomy: Navigating Unmanned Construction

Integrated at the core of the created autonomy for robotics in construction is the never-seen-before precision of monitor and control systems. Real-time information on construction sites that is collected by sensors and IoT devices that are installed on the construction sites enables the robotic system to complete its operations smoothly (Ammad et al., 2021). All this real-time information is then submitted to edge computing and cloud technologies, which enable autonomous systems to make optimal decisions for construction in related environments (Liang et al., 2024).

Robotic autonomy when applied to construction might revolutionize the sector for the better by tackling important difficulties including the shortage of manpower, safety concerns, and higher production. Each construction vehicle from trucks and excavators to robotic bricklaying machines, and drones for 3D on-site printing of dwelling constructions is already in deployment in numerous construction projects globally. When it comes to reasons for developing robotic autonomy, the most crucial one is the potential to enhance employee safety. Thus, by eliminating individuals from harmful operations, automated technologies allow for minimizing the incidence of catastrophes and accidents at construction sites. Also, these robotic devices can work in places that can be inaccessible to humans or harmful to staff.

Autonomous Construction Equipment

The construction industry is increasingly embracing the innovative impact of digital technology making the automation of construction equipment a prominent topic of discussion. These high-end robots consist of various sensors, artificial intelligence, and hard controllers to address problems and are among the most recent developments in construction management. Intelligent construction equipment refers to a wide variety of mechanically operated equipment or vehicles that can be employed in construction activities across numerous construction sites without manual management (Rasheed et al., 2024). These instances of autonomy involve self-driving trucks as well as excavators, robotic bricklaying, construction drones, and 3D-printing drones: all these are revolutionizing the construction industry by achieving higher levels of efficiency, safety, and accuracy (Zhou et al., 2024). The essence of autonomy is found at the center of an intelligent, digital HC system that employs upgraded sensors to deliver status information on the construction site. LIDAR, radar, and cameras are incorporated into an instant detection system for obstacles, terrain, and the maneuvering of humans and other equipment. This comprehensive situational awareness, backed by sophisticated AI computations, allows autonomous machines to comprehend construction sites, chart their courses, and accomplish operations on their own to a considerable extent. Another motive is to increase employee safety; many construction sites have been recognized to have a risky environment that might be mitigated with the usage of autonomous construction equipment (Saad et al., 2024). By eliminating human operators, such intrinsically dangerous jobs might be removed coupled with the many disasters that occur routinely on construction sites, therefore boosting safety at worksites.

Precision and Efficiency

The utilization of robotic autonomy in construction has reached new degrees of precision and efficiency, fundamentally transforming how projects are carried out and accomplished for construction organizations and their clients. One of the primary benefits of the use of autonomous construction equipment is lower human error. A lot of traditional construction work, especially in grading, excavation, and the positioning of material, requires the ability and experience of the human operator. Autonomous systems, on the other hand, are programmed to accomplish tasks with extraordinary precision and repeatability. With sensors and powerful algorithms, the robotic machines can measure crucial parameters and monitor them to enable work execution within tolerances adhering to the design specifications (Ali et al., 2024). The degree of accuracy with which robotic devices can complete tasks at hand is far greater compared to human beings. In earthmoving activities, the precision of an autonomous excavator and grader forms the ground perfectly, allowing ideal contours for drainage and site preparation. In masonry constructions, robotic bricklayers can lay bricks at an amazing pace, bettering the greatest human worker, while retaining top quality. Similarly, in infrastructure projects, autonomous paving equipment may pave asphalt or concrete with unequaled evenness to enhance long-term durability and performance (Lloyd et al., 2024).

Integration with Digital Twins

A digital twin in construction and robotics has an entwined and interconnected relationship to bring forth new potential for innovation and efficiency throughout the sector. Central to all these processes is the digital twin: a computerized model or digital representation of a physical facility or construction project, constantly updated with real-time input.

This permits the blending of real-time data from on-site sensors and monitoring systems with advanced 3D modeling simulations, to deliver a dynamic and updated picture of a construction site and its components. This close integration between the physical and digital realms allows autonomous construction equipment to navigate and operate on the construction site with exceptional precision and adaptability. The ability of the digital twin to foresee and simulate alternative scenarios can be exploited to direct the decision-making processes of the robotic system as it moves around the building site, maximizing the execution of jobs (Liu et al., 2024).

For example, a building site digital twin can model the terrain, barriers, and climatic conditions, enabling autonomous excavators and graders to accurately plan their motions and earth-moving tasks. Similarly, in infrastructure projects, the digital twin can imitate the structural behavior and performance of a bridge or tunnel, enabling autonomous inspection and maintenance equipment to monitor the asset's state and identify possible faults before they emerge in the actual world. This integrated nature goes beyond operational efficiency and adds to the overall safety and risk mitigation methods in construction projects. It will connect real-time data being collected by the autonomous systems with the predictive capabilities of the digital twin to enable construction teams to anticipate and minimize potential dangers, assuring the safety of their crew and the integrity of their work. With digital transformation more established in the construction industry, the combination of robotic autonomy and digital twins will become more significant. Firms in construction that have mastered the synergy from this pairing will be on the front line of innovation, giving exceptional value to their clients while bringing change to the built environment (Park et al., 2024).

Digital Twins and Robotics Contribution to Green Construction

The digital twin significantly contributes to the development of Intelligent Green Buildings (IGBs). It enhances the management of physical and engineering assets throughout their life cycles, especially during the operation and maintenance phases. Digital twins help maintain dynamic information, support activities, and the creation of a comfortable environment by integrating various systems such as computerized maintenance management systems, computer-aided facilities management systems, building automation systems, and integrated workplace management systems. Digital twins aid in the realization of sustainable and intelligent urban environments, contributing to the social, economic, and environmental progress of smart cities. They are used in designing energy efficient buildings for other purposes, as they involve modeling and analyzing various possible conditions to determine the best ways of saving energy. Digital twins also help in monitoring assigned resources, cutting down on wastage, and optimizing use of materials (Zaland et al., 2024). Digital twins are used in predicting problems that may arise in building systems and implementing solutions before the problems arise, hence reducing the time taken in maintenance, thus improving efficiency. Digital twins enable an integration of various professionals such as architects, engineers, and facility managers since they form a central platform for data sharing.

Despite their potential, the application of digital twins in IGBs is not yet fully integrated into the design process. There is a need for forward-looking designs that prioritize sustainable development, people's livelihoods, and green structures (Rasheed et al., 2023). Successful integration of digital twins into IGBs requires correct execution of the basic building processes and an effective system architecture to manage information distributed across multiple databases.

Robotics in Green Construction

In green construction practices, robotics are pivotal due to their capacity to enhance efficiency and precision. By executing repetitive tasks, it contributes significantly to the design, development, and construction processes, thus minimizing potential human errors. Automation in green construction relates to the use of machines and robots in enhancing construction activities (Aghimien et al., 2019). Thereby the integration results into improved efficiency and accuracy and enhanced safety in construction projects hence facilitating construction sustainability.

In construction activities, robots can accomplish tasks with superior accuracy, contributing to improved quality and uniformity which is significant in green buildings. Also, automation aids in the reduction of material wastage through the efficient use of resources and the avoidance of mistakes made by people (Cheah et al., 2022). The involvement of robots in construction work enhances the safety and health of construction workers as they avoid performing hazardous activities likely to result in accidents. Automation in construction is a reality that shows that construction activities will be performed in a shorter time and this will minimize the time taken to complete construction projects hence minimizing the use of energy and carbondioxide emissions.

Meanwhile robotics helps improve energy efficiency in green buildings by optimizing resource use and minimizing waste. Automated systems and robots can perform tasks with greater accuracy, reducing the need for rework and ensuring that materials are used efficiently. Robots support sustainable construction practices by enabling the use of advanced materials and innovative building techniques. They can assist in the creation of structures that meet high standards of energy efficiency and environmental sustainability, contributing to the overall goals of green constructions.

Synergy Between Digital Twins and Robotics

In green constructions, the use of digital twins and robotics complement each other for optimum results. Digital twins provide accurate data, enabling robotics to perform tasks with precision based on it (Xu et al., 2021). This creates a closed-loop system where the digital twin keeps track and analyzes the construction process in real-time while robots modify their operations according to the feedback received for efficiency and sustainability. Construction activities can be made more adaptable and responsive through the use of robotics that can modify construction procedures and strategies depending on the data and analysis offered by digital twins. In the use of resources in construction projects, the digital twins and robotics are integrated into their operations in a manner that makes the accomplishment of efficiency and sustainability possible. Consequently, the utilization of digital twins and robotics can enhance various aspects of construction and minimize the adverse impacts of human interventions on the environment, which in return makes construction more sustainable. With the advancements in the application of digital twins and robotics, the use of the two in green constructions will continue to grow and encompass more stages in the construction process to offer better sustainability and efficiency.

Evolution of Digital Twin Cities

Digital twin cities have evolved through several stages, from 3D models to perceptive and predictive types. The progression involves incorporating spatial static models, sensing data, and dynamic social data to achieve comprehensive urban simulations and intelligent control. The core of digital twin cities is the urban information model, which relies on multi-source data fusion and intelligent facilities. The integration of digital twins and IGBs is essential for achieving energy conservation, emission reduction, and sustainable urban management. Digital transformation facilitates urban governance reforms and promotes the green transformation of cities. Data quality and collection methods are critical for effective urban smart management and monitoring carbon-dioxide emissions.

References

Aghimien, D. O., Aigbavboa, C. O., Oke, A. E. and Thwala, W. D. (2019). Mapping out research focus for robotics and automation research in construction-related studies. *Journal of Engineering, Design and Technology*, 18(5): 1063–1079. https://doi.org/10.1108/JEDT-09-2019-0237.

Ali, Z., Saad, S., Rasheed, K. and Ammad, S. (2024). AI future perspectives and trends in construction. In *AI in Material Science* (pp. 239–261). CRC Press. https://doi.org/10.1201/9781003438489-11.

Ammad, S., Saad, S., Bashir, M. T., Qureshi, A. H., Altaf, M. et al. (2021). Construction accidents via integrating Building Information Modelling (BIM) with emerging digital technologies a review. *2021 3rd International Sustainability and Resilience Conference: Climate Change*. https://doi.org/10.1109/IEEECONF53624.2021.9668066.

Bay, C. (2023, June 27). *Powering Microsoft Smart Buildings with Microsoft Azure Digital Twins*. https://www.microsoft.com/insidetrack/blog/powering-microsoft-smart-buildings-with-microsoft-azure-digital-twins/.

Bhattacharyya, N. (2024). Role of data analytics and advanced computing in resolving selectivity and sensitivity issues in thick or thin film multi-sensor systems. In *Comprehensive Materials Processing* (pp. 525–535). Elsevier. https://doi.org/10.1016/B978-0-323-96020-5.00202-8.

Bilal, M., Iftikhar Ali, M., Rasheed, K. and Saad, S. (2023). Exploring adoption trends of knowledge management systems in the Software Industry's SMEs: A comprehensive investigation of enablers and barriers. *2023 4th International Conference on Data Analytics for Business and Industry (ICDABI)*, 598–604. https://doi.org/10.1109/ICDABI60145.2023.10629373.

Cardoso, J. L. da F. P., Rebentisch, E. S., Rhodes, D. H. and Soares, A. L. (2022). Value of concept of operations analysis for digital transformation using digital twins. *Product Management & Development*, 20(2): e20220006. https://doi.org/10.4322/pmd.2022.023.

Cheah, C. G., Chia, W. Y., Lai, S. F., Chew, K. W., Chia, S. R. et al. (2022). Innovation designs of industry 4.0 based solid waste management: Machinery and digital circular economy. *Environmental Research*, 213: 113619. https://doi.org/10.1016/j.envres.2022.113619.

Chen, S. -Q. and Bai, J. (2023). Data-driven decision-making model for determining the number of volunteers required in typhoon disasters. *Journal of Safety Science and Resilience*, 4(3): 229–240. https://doi.org/10.1016/j.jnlssr.2023.03.001.

Cheng, J. P., Wong, P. K. -Y., Luo, H., Wang, M., Leung, P. H. et al. (2022). Vision-based monitoring of site safety compliance based on worker re-identification and personal protective equipment classification. *Automation in Construction*, 139: 104312. https://doi.org/10.1016/j.autcon.2022.104312.

De Man, A., Uyttersprot, J. -S., Chavez, P. -F., Vandenbroucke, F., Bovart, F. et al. (2023). The application of near-infrared spatially resolved spectroscopy in scope of achieving continuous real-time quality monitoring and control of tablets with challenging dimensions. *International Journal of Pharmaceutics*, 641: 123064. https://doi.org/10.1016/j.ijpharm.2023.123064.

De Valence, G. (2022). *Creative Destruction and Constructing the Built Environment: From the First Industrial Revolution to the Fourth*. Construction Economics Research.

Demirdöğen, G., Diren, N. S., Aladağ, H. and Işık, Z. (2021). Lean based maturity framework integrating value, BIM and big data analytics: Evidence from AEC industry. *Sustainability*, 13(18): 10029. https://doi.org/10.3390/su131810029.

El-khawaldeh, R., Guy, M., Bork, F., Taherimakhsousi, N., Jones, K. N. et al. (2024). Keeping an "eye" on the experiment: computer vision for real-time monitoring and control. *Chemical Science*, 15(4): 1271–1282. https://doi.org/10.1039/D3SC05491H.

Grieves, M. W. (2023). *Digital Twins: Past, Present, and Future BT—The Digital Twin*. pp. 97–121. *In*: Crespi, N., Drobot, A. T. and Minerva, R. (eds.). Springer International Publishing. https://doi.org/10.1007/978-3-031-21343-4_4.

Haris, M., Saad, S., Ammad, S. and Rasheed, K. (2024). AI in fabrication and construction. In *AI in Material Science* (pp. 169–192). CRC Press. https://doi.org/10.1201/9781003438489-8.

Hirano, K., Sato, T. and Suzuki, N. (2024). Real-time prediction of material removal rate for advanced process control of chemical mechanical polishing. *CIRP Annals*. https://doi.org/10.1016/j.cirp.2024.04.028.

Hirsch-Kreinsen. (2016). "Industry 4.0" as Promising technology: Emergence, semantics and ambivalent character. *Soziologisches Arbeitspapier Nr. 48/2016*, 48: 33.

Javaid, M., Haleem, A., Singh, R. P., Suman, R., Gonzalez, E. S. et al. (2022). Understanding the adoption of Industry 4.0 technologies in improving environmental sustainability. *Sustainable Operations and Computers*, 3: 203–217. https://doi.org/10.1016/j.susoc.2022.01.008.

Javaid, M., Haleem, A. and Suman, R. (2023). Digital Twin applications toward Industry 4.0: A review. *Cognitive Robotics*, 3: 71–92. https://doi.org/10.1016/j.cogr.2023.04.003.

Jones, D., Snider, C., Nassehi, A., Yon, J., Hicks, B. et al. (2020). Characterising the Digital Twin: A systematic literature review. *CIRP Journal of Manufacturing Science and Technology*, 29: 36–52. https://doi.org/10.1016/j.cirpj.2020.02.002.

Khan, M. I., Khan, N., Saad, S. and Khan, M. W. (2024). Artificial Intelligence in construction materials. In *AI in Material Science* (pp. 47–64). CRC Press. https://doi.org/10.1201/9781003438489-3.

Liang, C. -J., Le, T. -H., Ham, Y., Mantha, B. R. K., Cheng, M. H., Lin, J. J. et al. (2024). Ethics of artificial intelligence and robotics in the architecture, engineering, and construction industry. *Automation in Construction*, 162: 105369. https://doi.org/10.1016/j.autcon.2024.105369.

Liu, L., Zeng, N., Liu, Y., Han, D., König, M. et al. (2024). Multi-domain data integration and management for enhancing service-oriented digital twin for infrastructure operation and maintenance. *Developments in the Built Environment*, 18: 100475. https://doi.org/10.1016/j.dibe.2024.100475.

Lloyd, S., Irani, R. A. and Ahmadi, M. (2024). Precision robotic deburring with simultaneous registration and machining for improved accuracy, quality, and efficiency. *Robotics and Computer-Integrated Manufacturing*, 88: 102733. https://doi.org/10.1016/j.rcim.2024.102733.

MarketsandMarkets. (2024, March). *Digital Twin Market Size, Share, Industry Report, Revenue Trends and Growth Drivers*. https://www.marketsandmarkets.com/Market-Reports/digital-twin-market-225269522.html.

Mazumder, A., Sahed, M. F., Tasneem, Z., Das, P., Badal, F. R. et al. (2023). Towards next generation digital twin in robotics: Trends, scopes, challenges, and future. *Heliyon*, 9(2): e13359. https://doi.org/10.1016/j.heliyon.2023.e13359.

Mihai, S., Yaqoob, M., Hung, D. V., Davis, W., Towakel, P. et al. (2022). Digital Twins: A survey on enabling technologies, challenges, trends and future prospects. *IEEE Communications Surveys & Tutorials*, 24(4): 2255–2291. https://doi.org/10.1109/COMST.2022.3208773.

Nardo, M., Forino, D. and Murino, T. (2020). The evolution of man–machine interaction: The role of human in Industry 4.0 paradigm. *Production & Manufacturing Research*, 8(1): 20–34. https://doi.org/10.1080/21693277.2020.1737592.

Nessari, S., Ghanavati-Nejad, M., Jolai, F., Bozorgi-Amiri, A., Rajabizadeh, S. et al. (2024). A data-driven decision-making approach for evaluating the projects according to resilience, circular economy and industry 4.0 dimension. *Engineering Applications of Artificial Intelligence*, 134: 108608. https://doi.org/10.1016/j.engappai.2024.108608.

Oliva, G., Cangialosi, F., Grimaldi, M., Fasolino, I., Belgiorno, V. et al. (2024). Urban odour annoyance management: An advanced embedded system for real-time monitoring enhanced by citizen science. *Case Studies in Chemical and Environmental Engineering,* 9: 100712. https://doi.org/10.1016/j.cscee.2024.100712.

Park, H., Shin, M., Choi, G., Sim, Y., Lee, J. et al. (2024). Integration of an exoskeleton robotic system into a digital twin for industrial manufacturing applications. *Robotics and Computer-Integrated Manufacturing,* 89: 102746. https://doi.org/10.1016/j.rcim.2024.102746.

Pei, S., Hoang, L., Fu, G. and Butler, D. (2024). Real-time control of urban drainage systems using neuro-evolution. *Journal of Environmental Management,* 360: 121133. https://doi.org/10.1016/j.jenvman.2024.121133.

Perno, M. and Hvam, L. (2020). *Developing a Framework for Scoping Digital Twins in the Process Manufacturing Industry.* https://doi.org/10.3233/ATDE200185.

Phillips, Z. (2020, November 11). *SoFi Stadium Digital Twin First of its kind | Construction Dive.* https://www.constructiondive.com/news/sofi-stadium-digital-twin-first-of-its-kind/588824/.

Presciuttini, A. and Portioli-Staudacher, A. (2024). Applications of IoT and advanced analytics for manufacturing operations: A systematic literature review. *Procedia Computer Science,* 232: 327–336. https://doi.org/10.1016/j.procs.2024.01.032.

Radich, A. S., Magalhães Filho, S. C., Borges, R. F. O., Scheid, C. M., Meleiro, L. A. C. et al. (2023). Real-time control and monitoring of drilling fluids density by fuzzy-based control system. *Geoenergy Science and Engineering,* 222: 211421. https://doi.org/10.1016/j.geoen.2023.211421.

Rane, S. B. and Narvel, Y. A. M. (2022). Data-driven decision making with Blockchain-IoT integrated architecture: A project resource management agility perspective of industry 4.0. *International Journal of System Assurance Engineering and Management,* 13(2): 1005–1023. https://doi.org/10.1007/s13198-021-01377-4.

Rasheed, K., Ammad, S., Said, A. Y., Balbehaith, M., Oad, V. K. et al. (2022). Application of smart IoT technology in project management scenarios. *2022 International Conference on Data Analytics for Business and Industry, ICDABI 2022.* https://doi.org/10.1109/ICDABI56818.2022.10041674.

Rasheed, K., Saad, S. and Ammad, S. (2023). Green financing for sustainable economic growth: Exploring critical factors towards resilient economies. *2023 International Conference on Sustainable Islamic Business and Finance (SIBF),* 104–111. https://doi.org/10.1109/SIBF60067.2023.10379984.

Rasheed, K., Saad, S., Ammad, S. and Bashir, M. T. (2024). Industry 4.0 and construction. In *AI in Material Science* (pp. 65–87). CRC Press. https://doi.org/10.1201/9781003438489-4.

Rasheed, K., Waqas, M., Saad, S., Zaland, A., Ali, Z. et al. (2024). Ethical and societal implications of AI in construction. In *AI in Material Science* (pp. 216–238). CRC Press. https://doi.org/10.1201/9781003438489-10.

Rasheed, K., Zaland, A., Saad, S., Ammad, S., Rostami, A. et al. (2024). History of AI. In *AI in Material Science* (pp. 15–46). CRC Press. https://doi.org/10.1201/9781003438489-2.

Saad, S., Haris, M., Ammad, S. and Rasheed, K. (2024). AI-assisted building design. In *AI in Material Science* (pp. 143–168). CRC Press. https://doi.org/10.1201/9781003438489-7.

Saad, S., Rasheed, K., Ammad, S., Ali, Z., Zaland, A. et al. (2024). The conclusive role of AI in Industry 4.0 construction materials. In *AI in Material Science* (pp. 262–270). CRC Press. https://doi.org/10.1201/9781003438489-12.

Saad, S., Rasheed, K., Ammad, S., Waqas Khan, M., Zaland, A. et al. (2024). Is AI the architect of tomorrow's materials in the age of Industry 4.0? In *AI in Material Science* (pp. 1–14). CRC Press. https://doi.org/10.1201/9781003438489-1.

Saad, S., Zaland, A., Ammad, S. and Rasheed, K. (2024). Robotics in construction. In *AI in Material Science* (pp. 115–142). CRC Press. https://doi.org/10.1201/9781003438489-6.

Sarker, D. (2023). *Achieving Digital Transformation Success: Interplay of Performance, Business Models and Digital Twins in Supply Chain Management.*

Sharma, A., Kosasih, E., Zhang, J., Brintrup, A., Calinescu, A. et al. (2022). Digital Twins: State of the art theory and practice, challenges, and open research questions. *Journal of Industrial Information Integration,* 30: 100383. https://doi.org/10.1016/j.jii.2022.100383.

Tang, Y., Huang, B., Wang, S., Zhou, J., Xiang, Z. et al. (2024). Computer vision-based real-time continuous monitoring of the pose for large-span bridge cable lifting structures. *Automation in Construction,* 162: 105383. https://doi.org/10.1016/j.autcon.2024.105383.

Waqas Khan, M., Saad, S., Ammad, S., Rasheed, K., Jamal, Q. et al. (2024). Smart infrastructure and AI. In *AI in Material Science* (pp. 193–215). CRC Press. https://doi.org/10.1201/9781003438489-9.

Wei, Y., Hu, T., Wang, Y., Wei, S., Luo, W. et al. (2022). Implementation strategy of physical entity for manufacturing system digital twin. *Robotics and Computer-Integrated Manufacturing,* 73: 102259. https://doi.org/10.1016/j.rcim.2021.102259.

Xu, W., Cui, J., Li, L., Yao, B., Tian, S. et al. (2021). Digital twin-based industrial cloud robotics: Framework, control approach and implementation. *Journal of Manufacturing Systems,* 58: 196–209. https://doi.org/10.1016/j.jmsy.2020.07.013.

Zaland, A., Saad, S., Ahmad, W., Ammad, S., Khan, M. W. et al. (2024). Enhancing mechanical properties of concrete by adding polypropylene fibers. *2024 ASU International Conference in Emerging Technologies for Sustainability and Intelligent Systems (ICETSIS),* 771–778. https://doi.org/10.1109/ICETSIS61505.2024.10459663.

Zaland, A., Saad, S., Rasheed, K. and Ammad, S. (2024). Automation in construction industry. In *AI in Material Science* (pp. 88–114). CRC Press. https://doi.org/10.1201/9781003438489-5.

Zhang, L., Guo, J., Fu, X., Tiong, R. L. K., Zhang, P. et al. (2024). Digital twin enabled real-time advanced control of TBM operation using deep learning methods. *Automation in Construction,* 158: 105240. https://doi.org/10.1016/j.autcon.2023.105240.

Zhou, Z., Abdi, E., Chea, C. P. and Bai, Y. (2024). Global path planning for autonomous construction vehicles in building construction: A comparative study with a focus on vehicle kinematic characteristics. *Journal of Building Engineering,* 109837. https://doi.org/10.1016/j.jobe.2024.109837.

Chapter 8

AI and ML Applications in Digital Twins and Robotics

Zawar Ali,[1] *Mohsin Ali,*[2] *Li Chen,*[2] *Kumeel Rasheed,*[1]
Syed Saad[3,*] and *Syed Ammad*[4]

Overview

The combination of robots and digital twins with artificial intelligence (AI) and machine learning (ML) have led to a transformational revolution in the construction industry in recent years. These technologies, now pivotal instruments, are transforming how we plan, construct, and interact with the built environment. The chapter embarks on an engaging exploration of AI and ML applications within robotics and digital twins, unveiling their practical implementations and scrutinizing their current utilization in the industry. In the construction industry, AI and ML have become essential, opening up previously unthinkable opportunities for increased productivity, cost reduction, and better project outcomes. When integrated with AI and ML algorithms, digital twin's virtual replicas of physical assets enable predictive maintenance, real-time monitoring, and performance optimization. Construction experts can simulate and analyze different situations, spot possible problems, and make well-informed decisions to increase productivity and lower risks by building a digital twin (Zafar et al., 2024).

[1] Department of Civil Engineering, CECOS University of IT and Emerging Sciences, Pakistan.
[2] School of Civil Engineering, Southeast University, Nanjing, China.
[3] Science, Engineering & Manufacturing Centre Sandwell College West Bromwich, United Kingdom.
[4] School of Engineering and Built Environment, Griffith University, Gold Coast, Australia.
* Corresponding author: syed.saad@sandwell.ac.uk

Furthermore, automation, accuracy, and efficiency in building operations have been made possible by robotics combined with AI and ML technologies. Robots possessing advanced sensory and cognitive capacities can execute difficult operations with outstanding accuracy, reducing human faults and improving total efficiency. Collaborative robots, commonly known as cobots, enhance the capabilities of human workers, promoting a safer and more efficient work environment. AI and ML are already being utilized in construction operations in a variety of fascinating ways. Construction forecasting involves using predictive analytics, which is driven by AI and ML algorithms. This helps stakeholders by predicting project outcomes, identifying potential hazards, and allocating resources as efficiently as possible. The utilization of data propels decision-making, minimizes delays and boosts project productivity (Wang et al., 2024). Additionally, AI-driven methods are revolutionizing robotics and digital twins by AI-driven methods. Digital twins can mimic and optimize construction processes with AI, allowing stakeholders to examine and visualize many scenarios before execution. Better design, more efficient use of resources, and lower costs are their byproducts. Furthermore, robotics with AI capabilities allows for the accurate and effective completion of jobs including material handling, concrete pouring, and bricklaying. These developments in robotics improve worker safety and well-being in addition to production (Baratta et al., 2024).

The chapter examines real-world applications of AI and ML in robots and digital twins in the construction sector. By embracing innovation and leveraging the potential of these technologies, we can drive productivity, efficiency, and sustainability in building projects, by realizing their revolutionary potential.

This chapter will specifically address the following areas:

- AI and ML in construction
- Examining how AI and ML are used in the construction sector
- Forecasting construction using predictive analytics
- Analyzing how ML and AI enable precise project forecasting
- Presenting evidence of real-world predictive analytics applications
- AI-powered techniques for robotics and digital twins
- Robotic precision enhanced by AI
- Intelligent visualization and simulation

Artificial Intelligence and ML in Construction

The construction industry is undergoing a significant transformation with the integration of AI and ML, which have proven to be highly beneficial in enhancing various aspects of the construction process. These technologies can be utilized in intelligent project management systems to help project managers simplify and streamline complex scheduling and execution processes, making them more efficient. Additionally, ML systems can analyze big data to detect trends and provide accurate forecasts. Consequently, the amount of work being done can be measured more accurately, reliable risk management strategies can be implemented, and resource allocation can be maximized. For instance, AI and ML algorithms were employed in the design of procurement and construction activities for a building project in Singapore, resulting in a 20% reduction in project duration and a 40% decrease in project costs (Datta et al., 2024).

Actually, automated quality control is another aspect of AI and ML application in the construction sector. Delivering high-quality construction necessitates stringent control of quality processes, which is accomplished through the automation of operations using advanced intelligent and machine learning technologies (Saad et al., 2024). It is precisely these creative technologies that make anomalies, inconsistencies , and deviations from the design noticeable through data from sensors, cameras and other sources. For instance, a robotic vision system which detects flaws in real time was successfully deployed in a major infrastructure project in the USA. The outcome was a 70% decrease in rework thereby minimizing waste, increasing efficiency and lowering the project cost (Waqar, 2024).

AI and ML also have an important effect on monitoring and controlling the risk at all manufacturing levels. These technologies analyze real-time data from sensors, drones and dash cams to anticipate potential safety issues, identify possible hazards, and provide proactive warnings. Worker safety in construction companies can be ensured through accident prevention, reduced injury incidents and enhanced workplace safety measures. An Australian construction site that implemented an AI-based safety monitoring system exemplifies this. The technology can detect potential safety risks such as worker falls, unauthorized access, and equipment malfunctions by analyzing video feeds from several cameras. Consequently, the site experienced a 50% reduction in accidents and enhanced overall security overall (Alwahedi et al., 2024). The construction industry is now leveraging AI and ML beyond just project management, quality assurance, and safety. By employing these technologies, virtual design and construction, create digital representations of building projects, enabling stakeholders to visualize and refine plans before construction begins (Haris et al., 2024). This approach enhances overall project productivity, reduces rework, and fosters better collaboration.

Intelligent Project Management Systems

Recently, project management systems powered by advanced AI and ML algorithms have not only transformed construction project planning but also revolutionized the entire execution process, making it unparalleled. These models can analyze vast amounts of data, such as completed project information, weather patterns and resource availability. This analysis aids project managers in making predictive decisions regarding scheduling, risk management, and resource allocation by identifying patterns and enabling the system to learn from past experiences. Recent statistics indicate a remarkable growth trajectory for AI in project management, with the global market expected to expand from USD 2.5 billion in 2023 to an estimated USD 5.7 billion by 2028, reflecting a substantial compound annual growth rate (CAGR) of 17.3% during the forecast period (LeewayHertz, 2024). The project's duration, cost, and resource utilization are the three main variables influenced by AI adoption as shown in Figure 1. The intelligent project management system's advantages include:

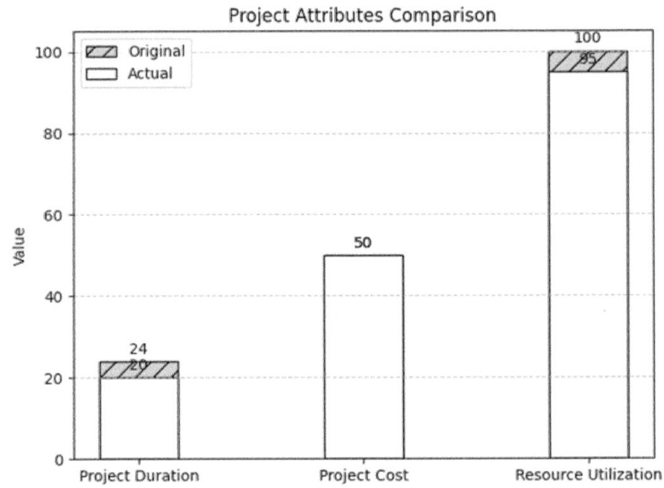

Figure 1: Comparison of project duration, cost, and resource utilisation.

Automated Quality Control

The advancement of AI and ML has transformed quality control systems in the construction industry. However, AI and ML, have now taken the lead in redefining quality control within this domain (Shang et al., 2023). These systems gather data on faults, non-conformance, and geometric variations from various sources including sensors, cameras, and Internet of Things (IoT) devices (Rasheed et al., 2021). For instance, AI systems can detect inconsistencies in the make-up of concrete and motivate workers to execute the necessary corrective actions by using digitized data obtained from a concrete pouring activity. Construction organizations can achieve more productivity, lower error rates, and better project quality by automating quality control. This lowers the need for rework, improves client satisfaction, and conserves time and resources (Son et al., 2023).

Safety Monitoring and Risk Reduction

AI and ML have emerged as important tools for safety monitoring and risk reduction. In the construction sector, safety is a primary responsibility. These technologies are capable of understanding expected safety threats, predicting safety risks, and providing proactive alerts to construction crews by analyzing real-time data from sensors, drones, and connected devices (Waqas Khan et al., 2024). For instance, smart device data can be evaluated by AI algorithms to follow employees' vital signs and identify indicators of stress or fatigue. Proactive steps can be taken to ensure a safer working environment and prevent accidents by informing supervisors of potential safety issues. Moreover, real-time monitoring of building sites by AI-powered video analytics can spot risky activities, illegal entry, and equipment failures. AI and ML help to minimize project delays, improve safety procedures, and lower accident rates by providing real-time monitoring and action (Zio and Miqueles, 2024).

Predictive Analytics in Construction Forecasting

In the field of construction forecasting, predictive analytics has become a disruptive force that is significantly transforming how the industry works. Construction specialists have now improved overall project management, optimized resource allocation, eliminated risks, and provided accurate forecasts regarding project outcomes by applying historical data, advanced statistical algorithms, and ML approaches. Predictive analytics has huge ramifications in the domain of construction forecasting. By evaluating historical project data, weather trends, and site conditions, ML approaches improve project planning and scheduling and help construction organizations plan ahead of time and allocate resources more effectively. Predictive analytics also greatly enhances cost estimation and budget management. Construction organizations provide precise cost estimates, minimize cost overruns, and enhance financial planning by examining historical cost data, market trends, and project-specific factors. Predictive analytics is also essential for safety precautions and risk reduction (Gifford and Bayrak, 2023). Predictive algorithms, which ensure a safer working environment for construction workers, can identify high-risk regions, predict safety events, and offer preventive measures by analyzing historical data and patterns.

The value and real-world applicability of predictive analytics in construction projects are evident from the occurrence of real-world events. Turner Construction Company, a North America-based, international construction firm has made notable strides in enhancing its business intelligence through the implementation of Project Pulse, a solution conceived from their innovation challenge initiative (Turner, 2023). By integrating data from platforms like SAP, Procore, Prolog, and Turner's data warehouse, Project Pulse enables real-time tracking and visualization of essential project metrics such as Change Management, RFIs, and Submittals. This tool has optimized the reporting process, providing real-time updates and customizable dashboards through Power BI, which can be quickly set up without adding extra work for staff. The adoption of project pulse has resulted in significant time savings, increased transparency, and improved client engagement by delivering easily understandable and consistently updated data.

Predictive Maintenance of Equipment

In the construction sector, tools are crucial for the proper execution of projects. Predictive analytics is being adopted by construction firms to improve operational efficiency and optimize equipment maintenance schedules. They use sensor data, past maintenance logs, and equipment performance indicators to identify possible equipment breakdowns before they occur by utilizing predictive analytics (Khan et al., 2024). By scheduling maintenance tasks at appropriate instants, businesses can minimize downtime and project schedule interruptions by taking a proactive approach. Construction organizations guarantee optimal equipment performance by taking preventive steps, such as replacing worn-out parts or performing maintenance operations, by anticipating future equipment faults. This reduces unscheduled downtime, boosts production, and significantly increases overall equipment efficiency (Ren et al., 2024). The essential elements of predictive equipment maintenance in the construction sector are displayed in Figure 2.

Material Demand Forecasting

For construction companies to manage their inventory effectively and run their supply chains more efficiently, accurate material demand forecasting is essential (Zaland et al., 2024). Predictive analytics now enables construction organizations to accurately forecast material

Figure 2: Components of predictive equipment maintenance.

demand. Construction organizations can avoid material shortages or excess supplies by managing their inventory levels and monitoring external factors such as supplier performance and market conditions, as well as prior consumption trends and project deadlines. Construction firms are able to make well-informed decisions about procurement and inventory replenishment by utilizing predictive analytics to detect patterns and trends in material consumption. They can optimize their material ordering processes to ensure that necessary resources are accessible when needed and minimize delays caused by shortages by understanding demand changes and matching them with project timetables (Maisel et al., 2023).

Construction businesses can react quickly to shifts in the market, such as changes in supplier availability or pricing, owing to predictive analytics (Saad et al., 2024). Businesses can modify their procurement tactics and find substitute suppliers or materials to reduce all possible hazards by adding outside variables into the predictive models. Further innovations and optimizations in the fields of advanced analytics and digital transformation have tremendous potential as the construction sector continues to embrace these trends. The flow of data and multiple elements implicated in material demand forecasting, alongside the

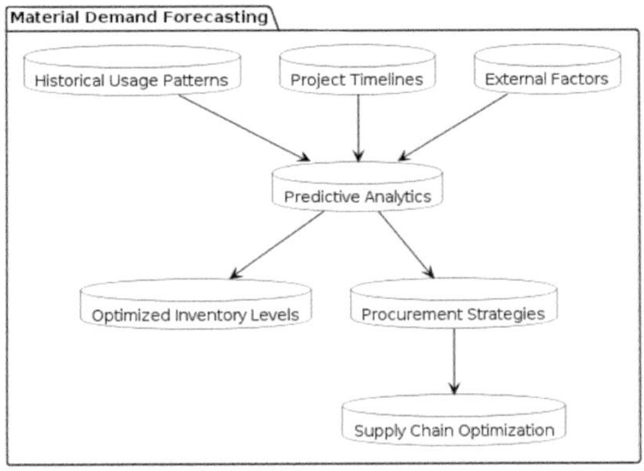

Figure 3: Material demand forecasting for optimized inventory management and efficient supply chain operations in the construction industry.

subsequent optimization of supply chain and inventory management, are crucial in the construction sector, as displayed in Figure 3.

AI-Driven Strategies in Digital Twins and Robotics

The AI-Driven methodology comprises employing AI algorithms and data-driven insights to maximize performance, improve decision-making processes, and stimulate improved automation across diverse industries. This aims to explore the importance, practical consequences, and present implementations of AI-driven strategies in the domains of digital twins and robotics. The practical implications of AI-driven strategies in digital twins and robotics are manifold across several industries. By integrating AI capabilities into digital twin models and robotic systems, organizations can achieve enhanced operational efficiency, reduce maintenance costs, enhanced safety, and increased production (Zafar et al., 2024). An AI-driven approach, a key aspect of smart cities, is closely linked to digital twins and AI-orientated robotics. This integration can facilitate popular route optimization, reduce energy consumption, and enhance town planning among other benefits.

- Aircraft manufacturers use AI-driven strategies in digital twins and robotics to optimize flight operations, enhance fuel efficiency, and improve maintenance procedures.
- Robotic systems integrated with AI technologies enable real-time imaging analysis, aiding surgeons in visualizing internal structures and making informed decisions.
- Companies analyze large volumes of historical data to identify trends, predict demand fluctuations, and optimize inventory management (Wang et al., 2024).

- AI algorithms optimize inventory levels, resulting in reduced costs, minimum stockouts, and improved customer satisfaction.
- AI-driven strategies in digital twins and robotics offer increased productivity, cost savings, improved decision-making, and enhanced safety.
- Integration of AI algorithms and data-driven insights enables proactive maintenance, optimized performance, and risk mitigation (Dutta and Kumar, 2024).
- Real-time data analysis and predictive modeling help organizations stay ahead in competitive markets.

There is no doubt that many changes and amendments are taking place in AI-driven strategies and will continue to do so in the near future. AI technology is here to stay, so the next focus is on how to merge AI with robotics for further enhancement. Nowadays, there are new technologies, such as optical technology and voice recognition, that are making robots smarter by allowing them to observe their environment and communicate with it. This, in addition, opens the gate for more sophisticated programs like driverless cars, cobots or those who can work alongside human beings, and even robotic systems that can adapt when conditions change (Rasheed et al., 2024).

AI-Driven Predictive Maintenance

AI-Powered Predictive Maintenance is a revolutionary method that applies AI algorithms and past data to predict equipment failures. This method enhances maintenance processes and reduces downtime particularly in digital twins and robotics. This theme has been broadly accepted by the business community since organizations recognize the possibility of ancillary expenditure reduction, device performance enhancement, and extended life cycle and availability of equipment. Through ML algorithms derived from data including sensor readings, maintenance records, and equipment performance history, AI-driven predictive maintenance takes the task of data analysis upon itself on a large scale. On the other hand, these algorithms use such patterns as correlations or anomalies to develop predictive models capable of predicting when the device is likely to fail. The forward-thinking attribute of preventive maintenance enables businesses to accurately schedule regular inspection and maintenance activities in advance, thereby, reducing the risk of unplanned downtime that comes with sudden breakdowns (Stohr et al., 2024).

A significant example of AI-driven predictive maintenance can be found in the aerospace industry. Aircraft manufacturers have embraced digital twin models linked with AI systems to analyze real-time data from sensors integrated in aircraft. By continuously examining this data, AI systems can detect early indicators of equipment degradation or breakdowns, enabling maintenance crews to take prompt preventive measures. In addition, AI-driven predictive maintenance is crucial in vital industries such as energy and utilities, where equipment failures can have serious implications (Saad et al., 2024). By embedding AI algorithms in digital twin models, utility firms monitor the condition of electricity grids, pipelines, and other infrastructure. Early detection of anomalies or probable breakdowns enables for early repair and eliminates costly service interruptions.

AI-Enhanced Robotics

AI-enhanced robotics is a trend that combines AI techniques, such as computer vision, natural language processing, and ML algorithms, with robotics to enhance the capabilities of robots. This integration allows robots to perceive and interact with their environment more intelligently, leading to substantial improvements in numerous industries. One significant application of AI-enhanced robotics is evident in the healthcare industry. Surgeons can now employ robotic devices integrated with AI technology to execute complex surgeries with greater precision and accuracy. These robotic devices apply computer vision algorithms to interpret real-time imaging data, providing surgeons with detailed visual knowledge about internal structures. By adding AI algorithms, robots can assist surgeons in making informed decisions during surgical procedures, leading to improved patient outcomes and a better quality of healthcare delivery (Lin et al., 2023).

In addition, using the power of AI robots can make processes more efficient and safer. Machines can be trained by adopting the ML algorithms to get certain jobs done autonomously or in teams with human operators (Naeem et al., 2022). This allows robots to adapt to various environments and modify their actions based on real-time feedback, ultimately, this elevates automation to a higher performance level. While the healthcare industry is not the only one to leverage AI-integrated robots, others such as manufacturing, logistics and agriculture also can profit from the technology. The area of production is no longer exclusively the territory of humans when robots become their "colleagues" and are now programmed with computer vision skills, allowing them to precisely assemble all the delicate components without fail, thus adding more to production efficiency and product quality. In logistics, robots with the learning algorithms can drive themselves through the warehouse's complicated pathways, flexibly managing inventory, and effectively ordering processing. In agriculture, robots can exploit computer vision to locate and thus selectively harvest crops, hence reducing labor cost and enhancing agricultural productivity (Arkouli et al., 2021).

AI-Enhanced Robotic Precision

AI-enhanced robotics is utilized in construction quality control, alongside project planning and resource allocation. By employing computer vision and ML techniques, robots can examine construction components and detect any deviation from the prescribed design standards. By using this automated quality control procedure, construction is guaranteed to adhere to rigorous standards and minimize the likelihood of errors and rework, hence enhancing overall accuracy. Real-time monitoring and analysis using AI technologies are also gaining significance in construction (Saad et al., 2024). By incorporating sensors into construction equipment and structures, real-time data on structural integrity, ambient conditions, and equipment performance may be obtained. Subsequently, AI algorithms examine this data in real time, finding potential faults, forecasting failures, and notifying construction crews to implement preventative measures. This real-time monitoring and analysis assists in preserving construction precision, lowering the risk of structural breakdowns, and ensuring the safety of construction personnel. The practical deployment of AI-enhanced robotic precision in building is supported by various case studies. For example, Construction Robotics has developed the SAM (Semi-Automated Mason) bricklaying robot, which employs computer vision and ML algorithms to lay bricks with high precision and speed. At the basis of SAM's capabilities lies its comprehensive integration of computer vision and ML technologies, which aids in monitoring the environment, discovering variations, and executing bricklaying jobs with unmatched accuracy. This technology has been successfully implemented in different construction projects, resulting in enhanced productivity and improved construction quality (Habuza et al., 2021). Such real-time engagement ensures construction precision and reduces the risk of structural failures.

Automated Robotic Construction

In recent years, the construction sector has experienced an enormous increase in the deployment of AI-enhanced robots for diverse building activities (Ali et al., 2024). These advanced robotic systems come with the power to conduct difficult activities such as bricklaying, concrete pouring, and welding, demonstrating extraordinary levels of precision and efficiency. The adoption of AI-enhanced robots in building jobs indicates a substantial shift from conventional methodologies. These robots are outfitted with complex algorithms that incorporate them with the capacity to dynamically adjust to changing conditions (Xu et al., 2022). The incorporation of AI-enhanced robots into construction jobs offers various advantages that contribute to their expanding popularity. These robots exhibit the potential to operate in hazardous areas and conduct jobs that pose risks to human personnel. This not only promotes the safety of work sites but also enables the performance of demanding operations with unprecedented precision and speed. Furthermore, AI-enhanced robotic systems can operate constantly without succumbing to fatigue, resulting in shortened project completion times and greater productivity (Davila Delgado et al., 2019).

Moreover, the introduction of AI-enhanced robots promotes construction site safety by decreasing the exposure of human workers to dangerous situations, while also improving overall project schedules and customer satisfaction. The successful adoption of AI-enhanced robots in construction is evident through different real-world applications. Notably, the deployment of autonomous bricklaying robots has proven their capacity to lay bricks with extraordinary precision and speed (Zhao and Cheah, 2023). This technology has not only resulted in higher production but has also elevated construction quality by ensuring perfect alignment and minimizing errors. Figure 4 illustrates an in-depth overview of rising employment of AI-enhanced robotic systems in the construction sector, together with the related productivity benefits obtained over the years. The deployment of AI-enhanced robotic systems in the construction

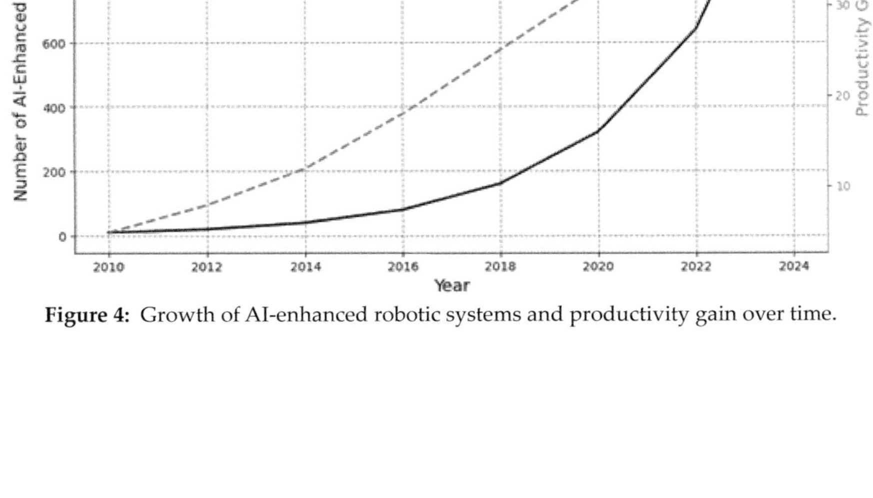

Figure 4: Growth of AI-enhanced robotic systems and productivity gain over time.

industry has experienced substantial growth, nearly doubling every two years.

Real-time Data Monitoring and Analysis

To identify bottlenecks, deviations from the project plan, and cost overruns, businesses employ real-time data measures such as project progress monitoring, resource allocation tracking, and cost performance analysis. This not only allows for a prompt response and measures needed to be taken to keep projects on schedule and costs under control, but also builds trust between consultants and their clients. Real-time data analysis benefits not only shareholders but also the entire project team, providing real-time information to support critical decision-making and communication (Chen et al., 2024). The implementation of real-time data monitoring and analysis utilizing AI and ML is transforming the construction business. The continuous stream of data from sensors integrated in construction equipment and structures, coupled with strong AI algorithms, enables proactive decision-making, early detection of issues, and optimization of resources. This technology has the potential to enhance construction safety, boost project efficiency, and optimize resource use (Barbier et al., 2024). As the industry continues to adopt real-time data monitoring and analysis, construction organizations harness the insights gathered to drive innovation, boost efficiency, and achieve sustainable growth. Figure 5 illustrates the real-time data monitoring and analysis trends in the construction industry.

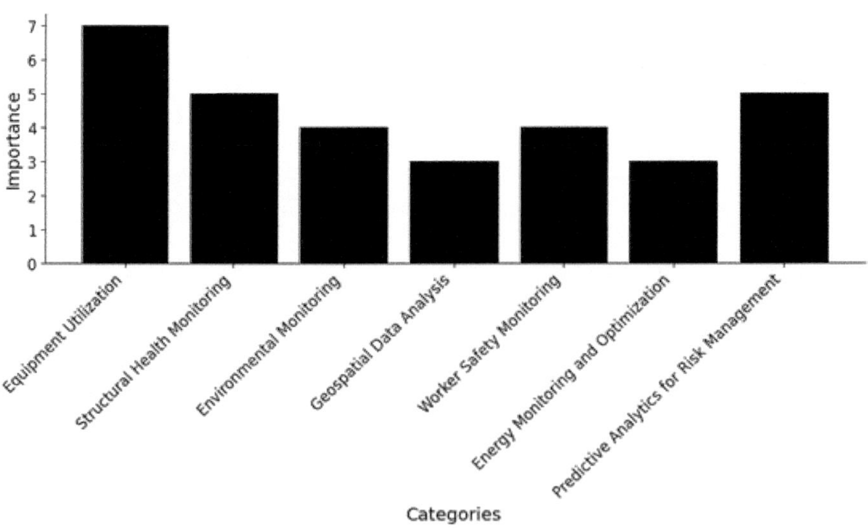

Figure 5: Importance of real-time data monitoring and analysis in construction industry.

Enabling Smart Simulation and Visualization

In recent years, driven by advancements in technologies and tools, smart simulation and visualization have emerged as a modern alternative approach that effectively generates virtual replicas and graphical depictions of construction projects through the application of mixed reality, virtual reality and augmented reality technologies and platforms. Such technology is particularly helpful as it enables better planning of projects, clearer communication among stakeholders and speedier decision conversion procedures (Rasheed et al., 2024). Furthermore, the capabilities of smart simulation and visualization provide an integrated platform, enabling key stakeholders to visually explore, demonstrate, and clarify the impact of decisions on the design within a simulated environment, ultimately facilitating informed decision-making. The construction sector is more likely to adapt to smart simulation and visualization technologies at any stage of a project after their initial release (Yaqoob et al., 2023). Besides that, virtual reality (VR) and augmented reality (AR) technologies have evolved with their mind-blowing implementations that support immersive experiences to the audience. The technologies employed in VR/AR allow all stakeholders to explore 3D virtual models of the project in real-time, fostering improved communication and mutual understanding among team members. As a result they work together more effectively and better decisions are made (Fleiß et al., 2024).

IoT devices capture real-time data from construction sites and input this information into virtual models, enabling dynamic simulations and analysis (Rasheed et al., 2022). This connection promotes decision-making by delivering precise and up-to-date information into project progress, resource consumption, and environmental conditions. It helps the building of precise digital models, enabling the simulation of numerous situations and the assessment of potential dangers. The usage of smart simulation and visualization tools optimizes project planning, improves communication among stakeholders, and boosts efficiency in decision-making processes. The integration of IoT devices and sensors, plus the deployment of AI algorithms, reflects growing developments that further expand the possibilities of smart simulation and visualization. By utilizing these technologies, the construction sector can achieve improved project outcomes, decreased costs, and enhanced collaboration among project stakeholders (Jararweh et al., 2023).

Building Information Modeling (BIM) and 3D Visualization

BIM signifies a collaborative method, which employs digital models to depict the physical and functional components of a building, a structure, a facility, or any infrastructure project. It increases the integration and coherence among the constructing elements, which range from architectural, to structural, to mechanical, using a virtual environment. This comprehensive advising system improves engagement, well-executed management, and high productivity during the course of project implementation. Additionally, the presence of 3D visualisation in BIM is equally important, as it enhances the methodology by offering viewers a more realistic and immersive depiction of the project in relation to the design. Using BIM 3D technology in real construction conditions has rapidly gained popularity during projects (Saad et al., 2023). One such engineering achievement was the birth of a massive trans-European rail project in London, known as the Cross-rail project. BIM, a vital tool for the project, has facilitated inspection and managed the complexities of various disciplines and stakeholders through its incorporation. Consequently, BIM has helped create a highly digital model that incorporates multiple systems, including railway, architectural, and structural components.

A prime example of the application and benefits of BIM and 3D visualization in planning and management is evident during the construction of skyscrapers in Shanghai, China. This renowned skyscraper significantly benefited from extensive use of BIM throughout all design and construction stages (Roche Cárcel, 2021). A BIM model consists of the joys of such complex structural systems, yet both work to boost building performance and enhance the depth of collaboration across various design and construction teams. This technology provides stakeholders the ability to physically see and study the building's appearance, lighting, and the different arrangements, which allows for the final design to be more accurate and efficient. Our BIM implementation in conjunction with 3D visualization helped significantly in reducing the time required for construction, errorless operation and achieving the final objective. However, following immediate needs there are those of the sector where visualization by BIM is the most popular at the moment. More and more AEC organisations tend to make BIM a standard part of their project pipelines, which demonstrates it being a huge part of the AEC workflow (Abuhussain et al., 2024). Among other BIM tools, software solutions like Autodesk Revit, Bentley Systems' AECOsim have proved to be priceless. However, the process of integrating BIM with other technologies, such as geographic information systems (GIS), laser scanning, and cloud computing, make BIM even more powerful and flexible, because the data flows become smoother, the precision increases and instant data sharing becomes possible.

Virtual Reality and Augmented Reality Applications

VR is the use of hardware that creates a virtual environment for the user, but AR displays digital information on top of the real environment. These technologies serve various purposes in the AEC business field, whether it results in a new, better or different way of conceiving, studying or developing projects (Al-Ansi et al., 2023; Sadamali Jayawardena et al., 2023). One of the most notable examples showcasing the practical application of VR and AR in the construction operations of the Qatar Foundation Stadium, a venue for the 2022 FIFA World Cup in Qatar, is the Qatar Stadium case. Using VR and AR technologies for visualization and optimization of the stadium design, the project allowed all stakeholders to participate in an interactive way which offered a real-time experience of the project. Through the designers' and engineers' use of virtual reality, they were able to test the stadium through a virtual version, and consequently gave insights on the spatial layout, the seat arrangements, and its aesthetic aspects. The experiential education gained by the design team throughout the project journey facilitated intelligent decision-making and allowed for adjustments and revisions to enhance the client and spectator experience as needed. AR was used as a tool during the construction process, allowing workers to overlay digital models on the physical site. This enabled them to accurately position building components and adhere to design criteria. These VR and AR techniques helped to speed up the design and construction process and resulted in an outstanding iconic stadium which also went well with the project objectives and also exceeded stakeholders' expectations.

The beneficial impact of VR and AR implementation in AEC is bolstered by concurrent advancements in technology and the increasing availability of user-friendly software. The prices of VR headsets and AR-compatible devices have significantly reduced and they are more user-friendly, thus making them integrate into the representative process workflow with ease. Such a solution is possible primarily in modern education. Software systems such as Unity3D, Autodesk's Forge, and Trimble's SketchUp Viewer can provide a plethora of advanced quality tools, for instance, the creation, viewing, and interaction with virtual and augmented realms.

Artificial Intelligence in Smart Simulation and Visualization

AI in smart modelling and visualization has evolved as a powerful tool across multiple fields, altering the way data is examined, patterns are found, and insights are generated. Smart simulation refers to the application of AI algorithms and techniques to construct realistic and interactive virtual worlds, whereas visualization involves the rendering of complicated data in a visual format. The incorporation of AI with smart simulation and visualization offers various benefits, including enhanced decision-making, improved understanding of complex systems, and increased efficiency. One interesting scenario illustrating the present utilization of AI in smart simulation and visualization is its application in the field of urban planning and design. Cities are growing increasingly complex, and planners have considerable hurdles in evaluating the influence of diverse design decisions on the urban environment (Sharifi et al., 2024). AI-powered smart simulation and visualization technologies can assist solve these difficulties by producing realistic digital models of cities and testing the impact of diverse initiatives. For example, AI systems can examine data on population density, traffic patterns, and energy use to model the impact of suggested improvements in transportation infrastructure or building energy efficiency (Badshah et al., 2024). Through interactive visualizations, planners can get insights into the potential results of diverse scenarios, enabling them to make informed decisions that increase urban sustainability, livability, and functionality (Ala et al., 2024).

AI in smart simulation and visualization extends beyond urban planning. While the field of medicine can make use of AI-driven simulation as a way of formulating individualized treatment programs and predicting disease outcomes, the same applies to agriculture. For example, AI systems analyze patient data including medical images, gene sets, and clinical records to roughly simulate sickness progression and drug treatments. These simulations allow healthcare providers to examine multiple treatment choices and forecast their efficacy, providing tailored and data-driven care. Visualization tools further enhance the understanding of complicated medical data, helping healthcare workers detect patterns, abnormalities, and connections that may not be evident in raw data (Rodríguez-Gracia et al., 2023). This integration of AI in smart modelling and visualization contributes to improved patient outcomes, enhanced diagnosis accuracy, and more effective treatment techniques. Moreover, AI in smart simulation and visualization has found applications in manufacturing and industrial processes. AI devices can analyze centrally the large volumes of sensor data emerging from production equipment in search for abnormal results, signaling the imminent failure or optimizing the production processes. AI application simulations can involve building virtual replicas of a factory system. Engineers can use these digital twins to analyze changes in conditions and optimize factory efficiency. Visualization tools enable engineers to monitor real-time data from many sources, identify bottlenecks, and depict complex industrial processes. This integration of AI with smart simulation and visualization results in reduced downtime, enhanced product quality, and enhanced overall productivity (Baduge et al., 2022).

References

Abuhussain, M. A., Waqar, A., Khan, A. M., Othman, I., Alotaibi, B. S. et al. (2024). Integrating Building Information Modeling (BIM) for optimal lifecycle management of complex structures. *Structures*, 60: 105831. https://doi.org/10.1016/j.istruc.2023.105831.

Al-Ansi, A. M., Jaboob, M., Garad, A. and Al-Ansi, A. (2023). Analyzing augmented reality (AR) and virtual reality (VR) recent development in education. *Social Sciences & Humanities Open*, 8(1): 100532. https://doi.org/10.1016/j.ssaho.2023.100532.

Ala, A., Simic, V., Pamucar, D. and Bacanin, N. (2024). Enhancing patient information performance in internet of things-based smart healthcare system: Hybrid artificial intelligence and optimization approaches. *Engineering Applications of Artificial Intelligence*, 131: 107889. https://doi.org/10.1016/j.engappai.2024.107889.

Ali, Z., Saad, S., Rasheed, K. and Ammad, S. (2024). AI future perspectives and trends in construction. In *AI in Material Science* (pp. 239–261). CRC Press. https://doi.org/10.1201/9781003438489-11.

Alwahedi, F., Aldhaheri, A., Ferrag, M. A., Battah, A., Tihanyi, N. et al. (2024). Machine learning techniques for IoT security: Current research and future vision with generative AI and large language models. *Internet of Things and Cyber-Physical Systems*, 4: 167–185. https://doi.org/10.1016/j.iotcps.2023.12.003.

Arkouli, Z., Kokotinis, G., Michalos, G., Dimitropoulos, N., Makris, S. et al. (2021). AI-enhanced cooperating robots for reconfigurable manufacturing of large parts. *IFAC-PapersOnLine*, 54(1): 617–622. https://doi.org/10.1016/j.ifacol.2021.08.072.

Badshah, I., Saad, S., Ammad, S., Rasheed, K., Khan, M. W. et al. (2024). Enhancing sustainable mobility: Lane change strategies for efficient traffic flow at stationary bottlenecks. *2024 ASU International Conference in Emerging Technologies for Sustainability and Intelligent Systems (ICETSIS)*, 1197–1202. https://doi.org/10.1109/ICETSIS61505.2024.10459583.

Baduge, S. K., Thilakarathna, S., Perera, J. S., Arashpour, M., Sharafi, P. et al. (2022). Artificial intelligence and smart vision for building and construction 4.0: Machine and deep learning methods and applications. *Automation in Construction*, 141: 104440. https://doi.org/https://doi.org/10.1016/j.autcon.2022.104440.

Baratta, A., Cimino, A., Longo, F. and Nicoletti, L. (2024). Digital twin for human-robot collaboration enhancement in manufacturing systems: Literature review and direction for future developments. *Computers & Industrial Engineering*, 187: 109764. https://doi.org/10.1016/j.cie.2023.109764.

Barbier, A., Salavert, J. M., Palau, C. E. and Guardiola, C. (2024). Analysis of the Euro 7 on-board emissions monitoring concept with real-driving data. *Transportation Research Part D: Transport and Environment*, 127: 104062. https://doi.org/10.1016/j.trd.2024.104062.

Chen, X., Wang, Y., Wang, J., Bouferguene, A., Al-Hussein, M. et al. (2024). Vision-based real-time process monitoring and problem feedback for productivity-oriented analysis in off-site construction. *Automation in Construction,* 162: 105389. https://doi.org/10.1016/j.autcon.2024.105389.

Datta, S. D., Islam, M., Rahman Sobuz, M. H., Ahmed, S., Kar, M. et al. (2024). Artificial intelligence and machine learning applications in the project lifecycle of the construction industry: A comprehensive review. *Heliyon,* 10(5): e26888. https://doi.org/10.1016/j.heliyon.2024.e26888.

Davila Delgado, J. M., Oyedele, L., Ajayi, A., Akanbi, L., Akinade, O. et al. (2019). Robotics and automated systems in construction: Understanding industry-specific challenges for adoption. *Journal of Building Engineering,* 26: 100868. https://doi.org/https://doi.org/10.1016/j.jobe.2019.100868.

Dutta, G. and Kumar, D. R. (2024). Strategic digitalization in SMEs of developing economies: Digital twin driven engineering value chain for customer-centricity. *Procedia Computer Science,* 232: 2654–2669. https://doi.org/10.1016/j.procs.2024.02.083.

Fleiß, E., Hatzl, S. and Rauscher, J. (2024). Smart energy technology: A survey of adoption by individuals and the enabling potential of the technologies. *Technology in Society,* 76: 102409. https://doi.org/10.1016/j.techsoc.2023.102409.

Gifford, M. and Bayrak, T. (2023). A predictive analytics model for forecasting outcomes in the National Football League games using decision tree and logistic regression. *Decision Analytics Journal,* 8: 100296. https://doi.org/10.1016/j.dajour.2023.100296.

Habuza, T., Navaz, A. N., Hashim, F., Alnajjar, F., Zaki, N. et al. (2021). AI applications in robotics, diagnostic image analysis and precision medicine: Current limitations, future trends, guidelines on CAD systems for medicine. *Informatics in Medicine Unlocked,* 24: 100596. https://doi.org/10.1016/j.imu.2021.100596.

Haris, M., Saad, S., Ammad, S. and Rasheed, K. (2024). AI in fabrication and construction. In *AI in Material Science* (pp. 169–192). CRC Press. https://doi.org/10.1201/9781003438489-8.

Jararweh, Y., Fatima, S., Jarrah, M. and AlZu'bi, S. (2023). Smart and sustainable agriculture: Fundamentals, enabling technologies, and future directions. *Computers and Electrical Engineering,* 110: 108799. https://doi.org/10.1016/j.compeleceng.2023.108799.

Khan, M. I., Khan, N., Saad, S. and Khan, M. W. (2024). Artificial Intelligence in construction materials. In *AI in Material Science* (pp. 47–64). CRC Press. https://doi.org/10.1201/9781003438489-3.

LeewayHertz. (2024). *AI in Project Management: Use Cases, Benefits, Techniques, Solution and Implementation.* https://www.leewayhertz.com/ai-in-project-management/#What-is-project-management.

Lin, T. -H., Chang, C. -T., Yang, B. -H., Hung, C. -C., Wen, K. -W. et al. (2023). AI-powered shotcrete robot for enhancing structural integrity using ultra-high performance concrete and visual recognition. *Automation in Construction,* 155: 105038. https://doi.org/10.1016/j.autcon.2023.105038.

Maisel, F., Neef, C., Marscheider-Weidemann, F. and Nissen, N. F. (2023). A forecast on future raw material demand and recycling potential of lithium-ion batteries in electric vehicles. *Resources, Conservation and Recycling,* 192: 106920. https://doi.org/10.1016/j.resconrec.2023.106920.

Naeem, U. H., Bilal, M., Syed, F., Rasheed, K., Saad, S. et al. (2022). A Multilayer encryption model to protect healthcare data in cloud environment. *2022 International Conference on Data Analytics for Business and Industry, ICDABI 2022.* https://doi.org/10.1109/ICDABI56818.2022.10041708.

Rasheed, K., Ammad, S., Said, A. Y., Balbehaith, M., Oad, V. K. et al. (2022). Application of smart IoT technology in project management scenarios. *2022 International Conference on Data Analytics for Business and Industry, ICDABI 2022*. https://doi.org/10.1109/ICDABI56818.2022.10041674.

Rasheed, K., Saad, S., Shahzad, L., Ammad, S., Ali, A. et al. (2021). Application of gesture data recognition in a human-interactive leap motion sensor chair. *2021 International Conference on Data Analytics for Business and Industry, ICDABI 2021*. https://doi.org/10.1109/ICDABI53623.2021.9655819.

Rasheed, K., Waqas, M., Saad, S., Zaland, A., Ali, Z. et al. (2024). Ethical and societal implications of AI in construction. In *AI in Material Science* (pp. 216–238). CRC Press. https://doi.org/10.1201/9781003438489-10.

Rasheed, K., Zaland, A., Saad, S., Ammad, S., Rostami, A. et al. (2024). History of AI. In *AI in Material Science* (pp. 15–46). CRC Press. https://doi.org/10.1201/9781003438489-2.

Ren, J., Zhang, J., Wang, J. and Zhao, X. (2024). Recursive signal denoising method for predictive maintenance of equipment by using deep learning based temporal masking. *Computers & Industrial Engineering*, 188: 109921. https://doi.org/10.1016/j.cie.2024.109921.

Roche Cárcel, J. A. (2021). The spatialization of time and history in the skyscrapers of the twenty-first century in Shanghai. *City, Territory and Architecture*, 8(1): 7. https://doi.org/10.1186/s40410-021-00136-z.

Rodríguez-Gracia, D., Capobianco-Uriarte, M., de las, M., Terán-Yépez, E., Piedra-Fernández, J. A. et al. (2023). Review of artificial intelligence techniques in green/smart buildings. *Sustainable Computing: Informatics and Systems*, 38: 100861. https://doi.org/10.1016/j.suscom.2023.100861.

Saad, S., Ammad, S., Rasheed, K., Balbehaith, M., Qureshi, A. H. et al. (2023). Enhancing construction management in restoration through BIM-assisted photogrammetric 3D model accuracy assessment. *2023 International Conference on Sustaining Heritage: Innovative and Digital Approaches (ICSH)*, 100–103. https://doi.org/10.1109/ICSH57060.2023.10482845.

Saad, S., Haris, M., Ammad, S. and Rasheed, K. (2024). AI-assisted building design. In *AI in Material Science* (pp. 143–168). CRC Press. https://doi.org/10.1201/9781003438489-7.

Saad, S., Rasheed, K., Ammad, S., Ali, Z., Zaland, A. et al. (2024). The conclusive role of AI in Industry 4.0 construction materials. In *AI in Material Science* (pp. 262–270). CRC Press. https://doi.org/10.1201/9781003438489-12.

Saad, S., Rasheed, K., Ammad, S., Waqas Khan, M., Zaland, A et al. (2024). Is AI the architect of tomorrow's materials in the age of Industry 4.0? In *AI in Material Science* (pp. 1–14). CRC Press. https://doi.org/10.1201/9781003438489-1.

Saad, S., Zaland, A., Ammad, S. and Rasheed, K. (2024). Robotics in construction. In *AI in Material Science* (pp. 115–142). CRC Press. https://doi.org/10.1201/9781003438489-6.

Sadamali Jayawardena, N., Thaichon, P., Quach, S., Razzaq, A., Behl, A. et al. (2023). The persuasion effects of virtual reality (VR) and augmented reality (AR) video advertisements: A conceptual review. *Journal of Business Research*, 160: 113739. https://doi.org/10.1016/j.jbusres.2023.113739.

Shang, X., Zhao, F., Du, J., Li, R., Xia, L. et al. (2023). Development and clinical evaluation of an online automated quality control system for improving laboratory quality management. *Clinica Chimica Acta*, 541: 117240. https://doi.org/10.1016/j.cca.2023.117240.

Sharifi, A., Tarlani Beris, A., Sharifzadeh Javidi, A., Nouri, M. et al. (2024). Application of artificial intelligence in digital twin models for stormwater infrastructure systems in smart cities. *Advanced Engineering Informatics*, 61: 102485. https://doi.org/10.1016/j.aei.2024.102485.

Son, Y. J., Pope, Z. C. and Pantelic, J. (2023). Perceived air quality and satisfaction during implementation of an automated indoor air quality monitoring and control system. *Building and Environment*, 243: 110713. https://doi.org/10.1016/j.buildenv.2023.110713.

Stohr, A., Ollig, P., Keller, R. and Rieger, A. (2024). Generative mechanisms of AI implementation: A critical realist perspective on predictive maintenance. *Information and Organization*, 34(2): 100503. https://doi.org/10.1016/j.infoandorg.2024.100503.

Turner. (2023, April 11). *Project Pulse: Advancing Business Intelligence | Insights | Turner Construction Company.* https://www.turnerconstruction.com/insights/project-pulse-advancing-business-intelligence.

Wang, S., Zhang, J., Wang, P., Law, J., Calinescu, R. et al. (2024). A deep learning-enhanced Digital Twin framework for improving safety and reliability in human–robot collaborative manufacturing. *Robotics and Computer-Integrated Manufacturing*, 85: 102608. https://doi.org/10.1016/j.rcim.2023.102608.

Waqar, A. (2024). Intelligent decision support systems in construction engineering: An artificial intelligence and machine learning approaches. *Expert Systems with Applications*, 249: 123503. https://doi.org/10.1016/j.eswa.2024.123503.

Waqas Khan, M., Saad, S., Ammad, S., Rasheed, K., Jamal, Q. et al. (2024). Smart infrastructure and AI. In *AI in Material Science* (pp. 193–215). CRC Press. https://doi.org/10.1201/9781003438489-9.

Xu, W., Huang, S., Han, D., Zhang, Z., Gao, Y. et al. (2022). Toward automated construction: The design-to-printing workflow for a robotic *in-situ* 3D printed house. *Case Studies in Construction Materials*, 17: e01442. https://doi.org/10.1016/j.cscm.2022.e01442.

Yaqoob, I., Salah, K., Jayaraman, R. and Omar, M. (2023). Metaverse applications in smart cities: Enabling technologies, opportunities, challenges, and future directions. *Internet of Things*, 23: 100884. https://doi.org/10.1016/j.iot.2023.100884.

Zafar, M. H., Langås, E. F. and Sanfilippo, F. (2024). Exploring the synergies between collaborative robotics, digital twins, augmentation, and industry 5.0 for smart manufacturing: A state-of-the-art review. *Robotics and Computer-Integrated Manufacturing*, 89: 102769. https://doi.org/10.1016/j.rcim.2024.102769.

Zaland, A., Saad, S., Ahmad, W., Ammad, S., Khan, M. W. et al . (2024). Enhancing mechanical properties of concrete by adding polypropylene fibers. *2024 ASU International Conference in Emerging Technologies for Sustainability and Intelligent Systems (ICETSIS)*, 771–778. https://doi.org/10.1109/ICETSIS61505.2024.10459663.

Zhao, X. and Cheah, C. C. (2023). BIM-based indoor mobile robot initialization for construction automation using object detection. *Automation in Construction*, 146: 104647. https://doi.org/10.1016/j.autcon.2022.104647.

Zio, E. and Miqueles, L. (2024). Digital twins in safety analysis, risk assessment and emergency management. *Reliability Engineering & System Safety*, 246: 110040. https://doi.org/10.1016/j.ress.2024.110040.

Chapter 9

Navigating Ethical and Social Implications in Digital Twins and Robotics

Muhammad Haris,[1] *Syed Saad,*[2] *Kumeel Rasheed,*[1]
Syed Ammad[3,*] and *Vipin Kumar Oad*[4]

Introduction

The developments of digital twins and robotics are advancing at the present time almost in all fields, including the medical and manufacturing sectors. In the present context, the term "digital twin" signifies a virtual replica of tangible entities, which allows for real-time monitoring and simulation of actual living organisms. In the field of robotics, machines are progressively gaining autonomy and skill in executing tasks previously reserved for humans. However, this advancement brings significant ethical and social implications that require careful consideration.

The introduction of Personal Digital Twins (PDTs) into different parts of human life sparks a deep look into their ethical and societal effects. The emphasis is on transferring our cognitive abilities and decision-making processes into digital formats, which might eventually be reintegrated

[1] Department of Civil Engineering, CECOS University of IT and Emerging Sciences, Pakistan.
[2] Science, Engineering & Manufacturing Centre Sandwell College West Bromwich, United Kingdom.
[3] School of Engineering and Built Environment, Griffith University, Gold Coast, Australia.
[4] Faculty of Civil and Environmental Engineering, Gdańsk University of Technology, Gdańsk, Poland.
* Corresponding author: syed.ammad@griffithuni.edu.au

into humans through sophisticated brain-computer interfaces. This major change raises important ethical questions about the ownership, control, and protection of personal data, which includes intimate details of our life and health histories. Issues about privacy, consent, and data security become prominent as PDTs aim to represent an individual's entire existence and physical condition (De Kerckhove, 2021). The merging of lifelogging, data tracking, and advanced algorithms for human replication emphasizes the need to address the ethical challenges of PDTs. The rise of "digital unconsciousness," where vast amounts of data about individuals remain unrecognized, raises profound ethical issues. The potential use of this digital unconsciousness for economic, social, or manipulative purposes highlights the urgent need for strict laws and ethical guidelines to protect people's privacy and freedom (Politou et al., 2022).

Moreover, the integration of lifelogs, personal data, and machine learning into fully developed PDTs demands thorough ethical consideration regarding the use and interpretation of this data. It is crucial to ensure that collecting and using such data aligns with the individual's interests, values, and fundamental principles (De Kerckhove, 2021). Additionally, the potential combination of lifelogs, personal data, and machine learning into complete PDTs requires extensive ethical reflection on managing and understanding this vast data. Ensuring that the data's use and interpretation ethically align with the individual's values and principles is essential in this rapidly advancing technological field (Ammad et al., 2021).

To operate optimally, digital twins tend to rely on significant numbers of personal attributes. This data could encompass personal information such as an individual's health status, habits, and even daily schedules, among other things. Privacy concerns arise from the heightened risk of this data being used without consent or being applied in unethical ways (Politou et al., 2022). The other ethical issue is; who is responsible for the decisions made by the self-actuating robots? With robots getting more autonomy and decision-making, it becomes necessary to determine who is to blame for a robot's actions particularly on instances when the robot has caused harm or loss. Moreover, applying digital twins and robotics into the environment of the workforce has social factors for instance job loss. On the positive side, these technological tools can enhance the efficiency and productivity thus improving the efficacy of the economy; on the negative side, they pose a threat of displacing human beings from the labor market hence causing disruptions in the economy.

Furthermore, the concept of digital unconsciousness suggests that a great deal of undetected personal data exists in digital domains. Regarding data ownership and individual control over one's digital identity, this idea presents serious ethical questions. When personal data is acquired and maintained without individuals' consent or knowledge, they may have little to no control over how it is used, shared, or even altered. The likelihood of privacy violations, as well as unlawful use and exploitation, raises ethical concerns about this lack of ownership and control over personal data.

To address these ethical issues, it is crucial to establish solid legal restrictions or rules and ethical standards. These regulations and standards should aim to protect individuals' rights and uphold social norms, especially in the context of mapping with digital twins and deploying robotics. Legal constraints should guarantee that people own and manage their digital identities and provide precise rules for the collection, use, and sharing of their data. A person's privacy and autonomy should be respected while integrating robotics and digital twins through the use of ethical guidelines that encourage openness, permission, and accountability. By implementing these measures, we can safeguard people's rights and maintain societal trust in the growing fields of digital technology and robotics (De Kerckhove, 2021).

Therefore, it is possible to state that not only digital twins and robotics have great opportunities, but their ethical and social analysis can become a challenge. These issues encompass privacy, accountability, displaced jobs, and consent. Tackling these concerns is essential for the fair and constructive utilization of these technologies.

Ethical Considerations in Digital Twins and Robotics

Digital twins and robotics are rapidly advancing and being employed to transform various industries by creating digital replicas of physical processes and automating systems. However, these are innovations that come with important ethical issues too. This outlines the key teaching points of ethical principles, emphasizes real-life scenarios related to these principles, identifies the problems they present, and suggests possible solutions. Big data managed by digital twins includes a significant amount of customers' personal information raising concerns about privacy and data protection. Safeguarding this information from leaks and misuse is essential not only to people's confidence but also to ethical utilization (Ali et al., 2024). For instance, in the medical field, constructing patients' digital twins can greatly enhance the efficacy of the customized therapy by conducting simulations and making estimations of the results. But if a hacker gains access to sensitive data such as patients' medical histories and other personal details, this would lead to serious violations of the affected patients' privacy. This demonstrates that improving data security and privacy necessitates the implementation of advanced security layers and practices to safeguard information (De Kerckhove, 2021).

In the realm of digital twins, safeguarding individuals' consent and autonomy over their data usage is paramount. This ethical imperative is especially crucial in contexts such as smart cities, where residents' data is utilized to develop digital twins of urban environments. Without clear and informed consent mechanisms, there is a risk of misuse and unauthorized access to personal information, potentially compromising privacy and trust. (Politou et al., 2022) emphasize the need for robust frameworks that enable individuals to manage the collection, usage, and protection of their data within digital twin initiatives. This approach fosters ethical practices and accountability in urban development.

Bias and fairness are critical ethical concerns in the deployment of algorithms within digital twins and robotics. These technologies often inherit biases from their training data, which can result in unjust outcomes. For instance, in the context of autonomous vehicles utilizing digital twins for navigation, biased data might lead to discriminatory practices in ride-sharing services. This scenario highlights the importance of ensuring that algorithms are trained on diverse and representative datasets to mitigate biases and promote fairness in decision-making processes (De Kerckhove, 2021). Addressing these issues is essential to build trust in the reliability and ethical integrity of digital twin and robotic technologies across various applications. One of the major key success factors in the use of digital twins and robots is the aspect of accountability or transparency most especially in manufactured goods (Rasheed et al., 2024). Considerable emphasis should be placed on the accountability assigned to these technologies for the decisions they make. This necessity arises from the requirement to verify the authenticity of all processes involved, thereby instilling confidence among all stakeholders in the system. For example, in manufacturing industries, robots use digital twins of the real environment for improvement of the production process. It is made sure that the algorithms, which control the robots, are safe and reliable. Moreover, this approach not only aids in risk management but also enhances industrial operations' acceptance and integration of these technologies (Politou et al., 2022). Meanwhile the incorporation of artificial intelligence (AI) and robotics has triggered transformative shifts across diverse sectors like healthcare and entertainment, promising heightened productivity and an enhanced quality of life. However, this evolution introduces a complex ethical terrain, requiring a delicate equilibrium between innovative progression and ethical obligations (De Kerckhove, 2021).

Digital Unconscious: Moral Quandaries and Privacy Concerns

The concept of 'digital unconscious' refers to extensive amounts of data collected about individuals, often without their conscious awareness. This data, gathered through interactions on the web and other physical devices, includes details about daily routines, personal preferences, and indicators of health (Naeem et al., 2022). Ethical concerns emerge from the collection and utilization of data that constitutes the digital unconscious. For instance, firms might gather extensive information about users' activities and preferences related to advertisements for targeted advertising purposes, even without the users' explicit consent. This brings to the fore, the ethics of using such data for commercial purposes without the respective individuals' consent. Privacy is a major concern with digital unconsciousness too. For instance, wristbands that continuously monitor the users' heart rates or sleep quality collect personal data. If unauthorized individuals access this information it could lead to discrimination against policyholders in insurance or employment opportunities based on their health data. Devices like Fitbit and Apple watch gather data on physical activity and sleep patterns and they also share this information with third parties, for purposes such as analyzing exercise and sleep routines. Similarly, certain social media platforms track user activities and interactions to tailor content and advertisements, leading to potential data leaks and manipulation.

As PDTs move from concept to actual implementation, tackling these ethical dilemmas becomes crucial for safeguarding individual rights, upholding societal values, and maintaining ethical standards. This necessitates dialogue among stakeholders, lawmakers, and thought leaders to establish appropriate guidelines for the responsible use of these powerful new technologies. It is imperative to explore significant concepts across various fields, keeping technological advancements in mind, to ensure that these beneficial innovations align with societal values of ethics and morality (Rasheed et al., 2024). The ethical considerations in digital twins and robotics and their key aspects and Implications can be seen in Table 1.

Impact on the Human Workforce: Social Landscape

The integration of DTs and robotics into manufacturing systems, especially in Human-Robot Collaboration (HRC), is becoming crucial for reshaping the human workforce. This tailored integration is not only drawing attention from the scientific community but also sparking increased scholarly discussions and research publications. DTs provide a

Table 1: Ethical considerations in Digital Twins and robotics: Key aspects and implications.

Aspect	Description
Impact of AI and Robotics	The convergence of AI and robotics triggers transformative shifts in various sectors, offering potential productivity enhancements and an enriched quality of life.
Ethical Terrain	The evolution introduces a complex ethical terrain, necessitating a delicate balance between innovative progression and ethical responsibilities.
PDTs	PDTs symbolize a shift externalizing cognitive functions and decision-making processes into digital realms, prompting reflections on potential future integration via brain-computer interfaces.
Emergence of Digital Unconscious	The scenario where extensive digital information about individuals remains unknown to them poses a moral quandary, emphasizing the need for stringent rules safeguarding privacy, freedom, and human self-control.
Ethical Duties in Decision-Making Processes	The ethical responsibilities associated with employing digital twins in critical spheres like healthcare and personal choices prompt questions about fairness, accountability, and transparency in decision-making by computers.
Addressing Ethical Quandaries	Addressing ethical dilemmas becomes crucial in preserving individual rights, societal values, and upholding ethical standards, requiring discussions among stakeholders to responsibly use these technologies.
Evolution of Digital Twins and Robotics	Digital twins and robotics are continuously evolving, demanding a multidisciplinary understanding and alignment with societal values to ensure these innovations adhere to ethical principles.

sophisticated platform for lifelike simulations, allowing detailed testing and refinement of collaboration strategies. This platform holds potential for boosting productivity, enhancing safety measures, and improving quality control in manufacturing systems. The adoption of DTs in HRC has shown measurable enhancements across different industries, including faster cycle times, better ergonomic considerations, and increased satisfaction among operators involved in these systems.

DTs are integrated with BIM platforms to simulate construction processes and coordinate human-robot interactions on-site. For instance, Skanska USA implemented DTs to plan and optimize robotic construction methods for building complex structures like hospitals (Manifold et al., 2022). This integration has led to more efficient construction workflows, reduced project timelines, and improved worker safety by automating repetitive tasks. Companies like Katerra use DTs to optimize robotic processes in prefabricated construction. DTs simulate assembly line operations, allowing precise control over robotic arms for manufacturing building components off-site. This approach has streamlined production, minimized material waste, and ensured consistent quality across prefabricated modules (Haris et al., 2024). In large-scale projects, DTs enable real-time monitoring and control of construction activities using robotics. For example, the use of drones equipped with DT technology by companies like Skycatch allows automated site surveys and progress monitoring (LNC Prakash et al., 2023). This integration enhances project visibility, improves logistics planning, and enhances safety by identifying potential hazards before they occur.

These achievements highlight the inherent potential for positive and beneficial impacts on the human workforce in these areas. However, the widespread use of DTs in HRC could lead to significant changes in employment dynamics, raising valid concerns about potential job losses (Kim et al., 2017). This emphasizes the urgent need for comprehensive retraining programs and targeted efforts to enhance skills that can adapt to and support the evolving roles and responsibilities within an HRC framework. Therefore, it is not only essential but imperative to actively develop and articulate a range of policies and strategies designed to minimize potential disruptions. At the same time, this approach aims to ensure fair distribution and sharing of the various benefits that come from integrating DTs into broader societal domains.

Digital Twins in Human-Robot Collaboration within Manufacturing Systems

The use of DTs for HRC in manufacturing systems is a relatively new area of research. It focuses on safety protocols, detailed production planning methods, ergonomic considerations, and creating collaborative environments in simulated settings. Recent interest and academic efforts in this field, particularly influenced by the disruptive impact of the Covid-19 pandemic, suggest the significant potential of DTs in improving HRC. This reflects a substantial endeavor to tackle practical challenges faced in complex industrial operations (Kim et al., 2017). Thus, while the combination of DTs and robotics promises significant improvements in boosting productivity and strengthening safety in manufacturing systems, the various effects on the human workforce require careful consideration of its societal and employment impacts. This calls for an equal focus on developing and executing comprehensive strategies designed to minimize potential disruptions and challenges. At the same time, these strategies aim to ensure a fair distribution of the many benefits that arise from integrating these technologies across different segments of society. Some of the key insights from the integration of digital twins and robotics are highlighted in Table 2.

Table 2: Integration of Digital Twins and robotics in manufacturing systems.

Aspect	Insights
Use of DTs for HRC in Manufacturing Systems	Focuses on safety protocols
	Detailed production planning methods
	Considerations for ergonomics
	Creation of collaborative environments in simulations
Recent Academic Efforts	Increased interest and research efforts
	Influenced by the Covid-19 pandemic
	Potential of DTs in enhancing HRC
Societal and Employment Implications	Effects on the human workforce require careful consideration
	Need for comprehensive strategies to minimize disruptions and ensure equitable benefits distribution
DTs and Robotics in Manufacturing Systems	Promise of improving productivity
	Strengthening safety measures
	Addressing challenges in complex industrial operations

Digital Inclusion and Accessibility in Construction Technology

In the realm of construction technology, ensuring digital inclusion and accessibility is crucial. This acts as a gateway to ensuring equitable access to technological advancements within the industry. This inclusivity reaches out to diverse groups like older adults and individuals with disabilities, aiming to bridge the digital gap they often face (Alamelu, 2013). This discussion aims to highlight the broad vision of digital inclusion while carefully exploring the many challenges involved. A major focus is on creating customized training programs and resources tailored to the needs of older and disadvantaged groups. This ensures smooth integration into the digital world (Smith, 2015). It is vital to address ethical issues related to data privacy, especially as digital services and products expand. This requires navigating complex decisions to ensure ethical practices while leveraging the potential of inclusive digital technologies.

Furthermore, establishing community-oriented Information and Communication Technology (ICT) support centers is a fundamental strategy for enhancing digital inclusion. These centers play a crucial role, not only providing technical assistance but also acting as catalysts for promoting comprehensive engagement within both business and societal contexts. Their importance lies in developing a culture of technological empowerment and accessibility across diverse segments of society. Emphasizing the essential role of digital literacy in enabling broad societal participation, there is a critical need to address the challenges and opportunities associated with achieving digital inclusion (Smith, 2015). This involves assessing the current state of digital inclusion, identifying significant barriers, and exploring potential solutions. Specifically, it brings to light the considerable difficulties that older adults and individuals with disabilities face when utilizing ICTs. These challenges encompass learning difficulties, intricate design features, insufficient support infrastructure, and a general disinterest in accessibility features (Saad et al., 2024). Addressing the shortage of accessible ICT learning resources is crucial, highlighting the urgency of overcoming these barriers to create a digitally inclusive environment for everyone (Nasibulina, 2015).

Ethical Dilemmas in Digital Inclusion

Furthermore, creating digital services and products that include everyone raises ethical questions, especially about gathering data and privacy (Politou et al., 2022). This requires coordinated efforts and collaboration among various groups like businesses, governments, and educators. Addressing design challenges, strengthening support systems, and managing new ethical issues all require a wide range of solutions. These include using age-friendly design methods, investing more in accessibility, and setting up community centers that offer ICT help in social settings. The ongoing problem of the digital divide highlights the stark differences in internet access across different groups. Factors like internet availability, skill levels, and access to content contribute to this gap. This inequality affects various parts of society, including age groups, education levels, genders, social classes, ethnicities, and people with disabilities. Despite efforts to bridge this divide, persistent challenges make it difficult for marginalized groups to establish and maintain an online presence (Zaland et al., 2024).

Nevertheless, achieving digital inclusion brings significant benefits, not just for individuals but for society overall and a wide range of stakeholders, including governments, service providers, designers, facilitators of assistive technology (AT), voluntary organizations, and social movements (Alamelu, 2013). Ensuring universal access to digital tools in construction projects is crucial. By promoting fair and transparent job opportunities, this effort helps people with disabilities become part of the construction industry. The role of AT in enhancing accessibility, particularly in less developed countries, cannot be overstated. Overcoming challenges in creating affordable and appropriately sized versions of these technologies is essential to enabling individuals with disabilities to actively participate in important societal roles and economic activities.

Transparency and Accountability: Building Trust in Construction

In construction project management, both accountability and transparency are the core values that should be embraced as they help in enhancing the activities' credibility, productivity, and successful conclusion. These principles are crucial for fostering public confidence, preventing misconduct in the construction industry, and ensuring effective project delivery (Filgueiras, 2016).

In construction, transparency is implemented intentionally to bring in people or allow other people to have either complete or partial access to information or the process of decision making that takes place within the organization (Khan et al., 2024). This commitment enables stakeholders to understand the logic behind actions and decisions during the existence of the project. The concept of accountability plays a central role of creating awareness to those who have the mandate of overseeing construction projects, that they are answerable for their deeds. It creates the basis of assigning responsibility of shareholders for their actions, decisions, and conduct during the lifecycle of the project. Accountability in the construction sector provides a strict check and balance in a manner that if something has been done they have to face the consequences hence making the construction sector highly responsible and reliable (Filgueiras, 2016).

Highlighting transparency and accountability in construction projects is not just about ethical behavior; it significantly enhances construction processes and outcomes overall. Transparency reduces uncertainty and risks while improving the quality of built environments. Moreover, holding stakeholders accountable ensures meticulous attention to detail and dedication, leading to improved project management and risk mitigation strategies. The importance of transparency and accountability goes beyond immediate project success. It influences broader societal perceptions and builds trust in the construction sector. Transparent construction practices drive social progress by ensuring that built environments promote sustainability, safety, and community well-being (Filgueiras, 2016).

In the construction industry, accountability and transparency are crucial for achieving positive outcomes, not just as moral standards. By developing a culture that values openness and responsibility, construction projects can achieve immediate success while making lasting contributions to socio-economic and environmental landscapes. Thus, transparency is regarded as critical in the construction industry because it provides full visibility into project planning and project delivery, budget distribution, decision-making, and all other essential information.

This openness ensures that there is transparency in the operation of the project and a true picture of the project is always depicted including the achievements that have been made, the problems that have been encountered and how resources are utilized. It also enhances free flow of communication with the clients, the public, the regulating authorities as well as the other stakeholders in the projects. Transparency decreases the probability of corruption, ineffective management, and unnecessary spending on projects putting every stakeholder into a favorable position (Filgueiras, 2016).

Foundations of Transparency and Accountability

Accountability in the construction industry is equally important, requiring that all parties participating in a project take responsibility for their choices and actions. Which include

- Government representatives
- Project managers
- Architects
- Engineers
- Contractors

The importance of openness and accountability is significant in the construction sector because of its complex processes, wide range of players, and large financial outlays (Filgueiras, 2015). These values support the long-term viability of the environment, maintain public perception, assure regulatory compliance, and aid in the effective completion of building projects (Heimstädt and Dobusch, 2020). The construction industry is one of the most massive and intricate industries that involve large monetary commitments; however, it plays a key social role for communities and societies. Transparency and accountability are crucial for maintaining integrity, enhancing productivity, and achieving goals and objectives in construction projects. These principles are not merely business-technical; they ensure gaining people's trust, avoiding unethical practices, and boosting confidence at every stage of construction (Filgueiras, 2016).

In addition to the core elements of decision-making and information provision, this framework consists of the following aspects. A decision is made clearly followed by clear communication on the rationale, thus building confidence with the stakeholders including investors, regulatory and planning authorities, communities, and architects and builders. Effective communication and openness foster a positive attitude among all key participants enabling them to work harmoniously together. Thus, when stakeholders understand why a specific project was chosen, they change their focus from pure business transactions and embrace the commitment of all participants towards the projects' success. It minimizes conflicts of interpretation, disagreements and creates a climate favorable for the continuation of project work. Also, they reduce instances of unethical behavior since decisions have been made based on legal and ethical standards; and improve the image of the industry regarding integrity.

Navigating the Landscape of Modern Construction Practices

Clear and transparent processes within the construction industry have a substantial effect on public perception. They allow communities to understand and support crucial construction decisions. This openness promotes inclusivity and engagement, empowering communities to voice their opinions, actively participate in projects that impact their environment, and engage in meaningful discussions (Kim et al., 2017). Consequently, this cultivates a sense of ownership and enhances confidence in the construction process and all involved parties. In essence, accountability ensures that individuals and groups in construction are accountable for their actions (Heimstädt and Dobusch, 2020). This helps create a culture of taking responsibility seriously while following ethical rules properly. Together, these principles navigate

- The intricate web of construction endeavors
- Shaping their outcomes and sculpting perceptions
- Regulatory compliance
- The sustainability of the built environment

Transparency and accountability are intertwined ideas that serve as important foundations in the sector of modern construction activities as shown in Figure 1, playing an important part in creating trust and efficiency within the sector (Heimstädt and Dobusch, 2020).

This field, known for its complex networks of people involved in various activities, depends greatly on ensuring transparency and strong accountability mechanisms (Zaland et al., 2024). These measures are crucial for fostering relationships, maintaining ethical behavior, and creating trust in construction projects. The increasing focus on these core aspects not only tackles the challenges of managing and carrying out projects but also deals with the intricacies of making ethical decisions. This approach significantly influences how construction projects are conducted, emphasizing the importance of building trust and maintaining integrity.

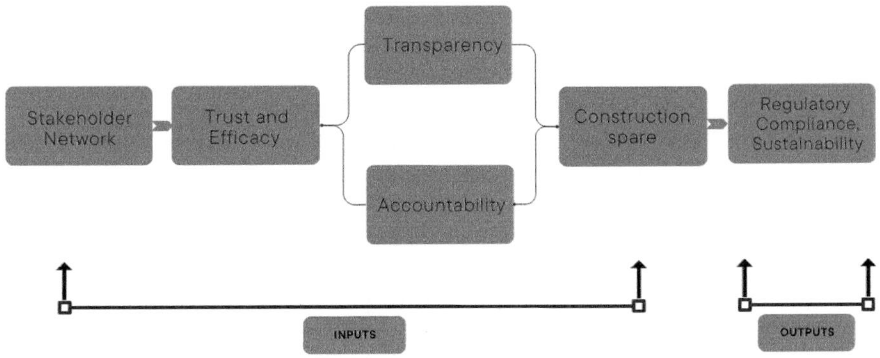

Figure 1: Interconnectedness of transparency and accountability in construction.

Environmental Ethics and Sustainable Practices

Environmental consciousness and ethical considerations are becoming more and more intertwined in the context of digital twins and robotics. Conversations on innovation are becoming more intelligent as a result of this changing narrative, which strikes a balance between the growth of science and a strong commitment to environmental preservation (Singh et al., 2019). Robotics and digital twins together provide an intriguing new frontier with many potential applications to transform operating environments in a variety of sectors. These state-of-the-art technologies offer improved efficacy, accuracy, and overall efficiency, and they tell a captivating story of development and advancement (Pater and Stadnicka, 2021).

However, amid this rapid technological progress, ethical dilemmas and sustainability issues emerge as significant considerations that require careful scrutiny. The convergence of these advanced technologies with heightened environmental awareness raises profound questions about the intricate web of resource allocation, waste management, and their broader impacts on the fragile balance of our natural ecosystems (Nasibulina, 2015). The incorporation of robotics and digital twins into various industries heralds a new era of innovation, yet it demands a thorough evaluation of its ecological footprint and ethical implications. As these technologies continue to advance, concerns about their potential impacts on resource consumption, waste generation, and environmental stability become increasingly pressing. Therefore, thoughtful deliberation on these matters is essential (Pater and Stadnicka, 2021).

Furthermore, the ethical compass guiding the deployment of robotics and digital twins necessitates a comprehensive analysis of their sustainable integration. It prompts careful consideration of how to implement these technologies to ensure that these technological advancements don't result in environmental deterioration. It becomes essential to strike a balance between environmental preservation and technology advancement in order to create innovative solutions that are both ground-breaking and morally sound (Al-Weshah et al., 2016).

With the mutually beneficial relationship between environmental concerns and technological discoveries growing, the need to balance growth with sustainability is becoming more imperative (Singh et al., 2019). The extent of the integration between robotics and digital twins necessitates a closer look at an ethical environment that not only entrusts us with our primary ecology but also continues to encourage technological development. As the digital landscape expands and robotics seamlessly integrate into everyday life, their environmental ramifications assume critical significance (Rasheed et al., 2024). This discourse endeavors to explore the myriad intersections of environmental ethics, sustainability imperatives, and the burgeoning field of robotics and digital twins, delving into their ethical dimensions and advocating for sustainable frameworks in technological advancement (Pater and Stadnicka, 2021). Through a multidisciplinary lens encompassing environmental philosophy, ethical considerations in technology, and sustainable methodologies, this investigation seeks to unveil pathways toward ethically sound and environmentally conscious advancements, steering the trajectory of robotics and digital twin applications towards a sustainable future. Robotics and digital twins have the ability to promote environmental ethics and sustainable practices in a number of ways as shown in Figure 2.

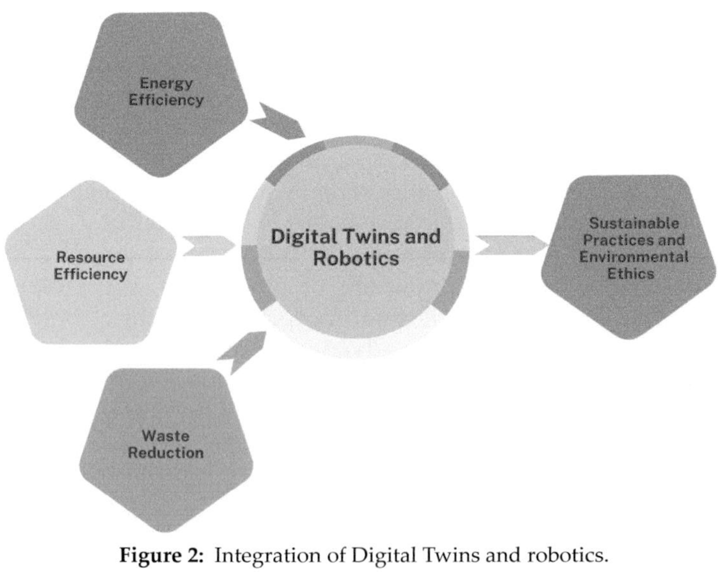

Figure 2: Integration of Digital Twins and robotics.

Strategic Integration of Digital Twins and Robotics for Sustainable Manufacturing Practices

Energy Efficiency

Energy efficiency in the field of digital twins and robotics is crucial in modern manufacturing. Combining digital twin technology with robots offers a way to optimize energy use and performance in factories (Pater and Stadnicka, 2021). This relationship between digital twins and robotics has the potential to transform manufacturing by blending physical and virtual spaces (Pater and Stadnicka, 2021).

Using digital twins in robotic systems helps bridge the gap between physical and virtual worlds. It enables real-time optimization, decision-making, and better energy management (Li et al., 2021). Despite much academic discussion on digital twin architecture, there's a lack of practical, experiment-based approaches to improve energy efficiency in robotic environments. This gap highlights the need for practical frameworks and solutions to optimize energy use and enhance efficiency in manufacturing. Exploring energy efficiency in digital twins and robotics is a promising area for advancing smart manufacturing. Integrating digital twins with robotics not only mimics physical objects but also improves understanding and efficiency in manufacturing processes (Pater and Stadnicka, 2021).

Through the establishment of interconnected layers like

- Physical infrastructure
- Digital models and
- Data exchange mechanisms

A comprehensive strategy unfolds, offering insights into energy-saving tactics, motion planning optimization, and operational enhancements.

Resource Efficiency

The emergence of Industry 4.0 indicates a transformation in the manner in which industries operate, driven by the integration of digital advancements and production processes (Saad et al., 2024). This revolutionary change redefines the basic concepts guiding manufacturing by emphasizing long-term, sustainable advancements over short-term trends. The merging of digital replicas and robotics is one example of how digital technology and industrial operations are crucial to this evolution (Waltersmann et al., 2021). The interaction between digital twins and robotics in Industry 4.0 is not merely a technological convergence; it represents a strategic partnership poised to revolutionize industrial processes (Saad et al., 2024). This convergence unlocks a new frontier of possibilities, paving the way for transformative advancements that signal a new era of manufacturing excellence (Domenech and Bahn-Walkowiak, 2019).

The core of this transformative potential lies in the synergy achieved by the harmonious merging of digital twins, accurately designed to mirror real-world entities or processes, and the precision-driven functionality of robotic systems. Digital twins, essentially digital replicas, act as virtual avatars of physical counterparts, offering a realm of possibilities to simulate, analyze, and optimize processes with unprecedented depth and accuracy (Pater and Stadnicka, 2021). But when these digital twins are seamlessly integrated with robotic system capabilities, manufacturing operations undergo a paradigm leap. This symbiosis expands the boundaries of resource optimization and reveals a world in which factories are flexible centers of production with effective resource allocation and use.

Understanding the path of Industry 4.0 reveals how important the combination of robots and digital twins will be in determining the direction of manufacturing in the future. The envisioned industrial environment is anchored by this mutually beneficial relationship, situated at the nexus of digital innovation and industrial efficiency (Saad et al., 2024). The integration of digital twins with robotic apparatuses charts a course towards unparalleled resource optimization, laying the foundation for factories that operate with precision and sustainability at their core. In this trajectory, the intricate interplay between digital twins and robotics is poised to transform conventional factory settings into smart, adaptive environments, wherein resource utilization becomes an orchestrated symphony of efficiency (Waqas Khan et al., 2024). The digital twins' capacity to replicate real-world entities or processes, coupled with robotics' precision and automation capabilities, heralds a new age where factories become agile entities, adept at optimizing resources, augmenting productivity, and championing sustainability in manufacturing practices (Domenech and Bahn-Walkowiak, 2019).

The profound implications of this convergence extend far beyond mere technological integration; it embodies a paradigmatic shift in the ethos of industrial operations. The fusion of digital twins and robotics redefines the very essence of factory operations, steering them towards an unprecedented realm of efficiency and sustainability (Omotayo et al., 2024). Such synergy enables manufacturing facilities to precisely manage resource allocation, ensuring optimal utilization while minimizing waste. Furthermore, it creates a revolutionary atmosphere where sustainability, speed, and accuracy come together to support an ecology that relies on flexibility and efficiency. Because of this, Industry 4.0's future path is largely dependent on the revolutionary potential that the combination of digital twins and robots will unleash, driving production to a level of sustainability and productivity never previously possible.

The relationship between robots and digital twins becomes crucial in determining how industrial evolution will go in the future as the story of Industry 4.0 develops. This strategic partnership signifies a fundamental change in the way that industrial processes operate rather than just a convergence of technologies (Domenech and Bahn-Walkowiak, 2019). The combination of digital twins' replicas that are precisely created with robotic systems' operational capabilities develops creativity, productivity, and sustainability in industrial environments. With the help of this collaboration, manufacturers will be able to operate outside conventional lines and create flexible, agile work environments with optimal resource use. The envisioned horizon of Industry 4.0 rests on the pillars of this symbiotic relationship, wherein digital twins and robotics become the driving forces behind a new era of manufacturing excellence, marked by resource efficiency and sustainable productivity.

Waste Reduction

The merging of digital twins and robotics in today's hi-tech world presents a promising but complex interaction. The integration of these fields effectively tackles the critical issue of managing and reducing waste in these sectors. This problem is especially evident in various industries like manufacturing, construction, and other industrial areas (Haris et al., 2024). Combining digital twins and robotics appears as a potential solution, offering a promising way to improve current waste management strategies (Chakraborty et al., 2022). Despite significant technological advancements, effectively managing waste in these dynamic hi-tech fields remains a complex and relatively unexplored area, requiring detailed research efforts and structured frameworks. These efforts aim not only to optimize resource use but also to reduce waste and enhance sustainability practices.

In these advanced industries, this new approach offers an opportunity to innovate and develop solutions that balance environmental care with technological progress, laying the foundation for a more sustainable future. The synergy between digital twins and robotics has the potential to bring about significant changes through focused research, innovative methods, and smart implementations, ensuring a more responsible and sustainable path for the future of tech-driven sectors (Chakraborty et al., 2022).

Digital Twin Implementations Supporting UN SDGs

Through enhanced efficiency, reduced waste, and optimized resource management across various sectors digital twins are already proving the ability to help achieve the UN-SDGs (United Nations Sustainable Development Goals). But what is critical at this point is conceptualization of digital twin technology scalability, as well as SDGs'. Digital twins are predominantly positioned in the manufacturing field while some entities are interested in expansion of the use in different areas such as urban planning, healthcare and education - the areas where SDGs can be resolved faster. Meanwhile Top tech companies like SpaceX utilize digital twins and robotics to enhance efficiency and safety in various aspects of space exploration, including spacecraft design, manufacturing, and operations (Arzo et al., 2023). Digital twins enable real-time monitoring and predictive maintenance of spacecraft components, while robotics streamline assembly processes and enable intricate tasks in space missions. However, digital twins appear to have a great potential of bringing a positive contribution to sustainable development and wider studies must be conducted to amplify the impact on SDGs (Allam and Jones, 2021). Meanwhile, the primary areas of digital twin applications align with the relevant SDGs, as illustrated in Figure 3.

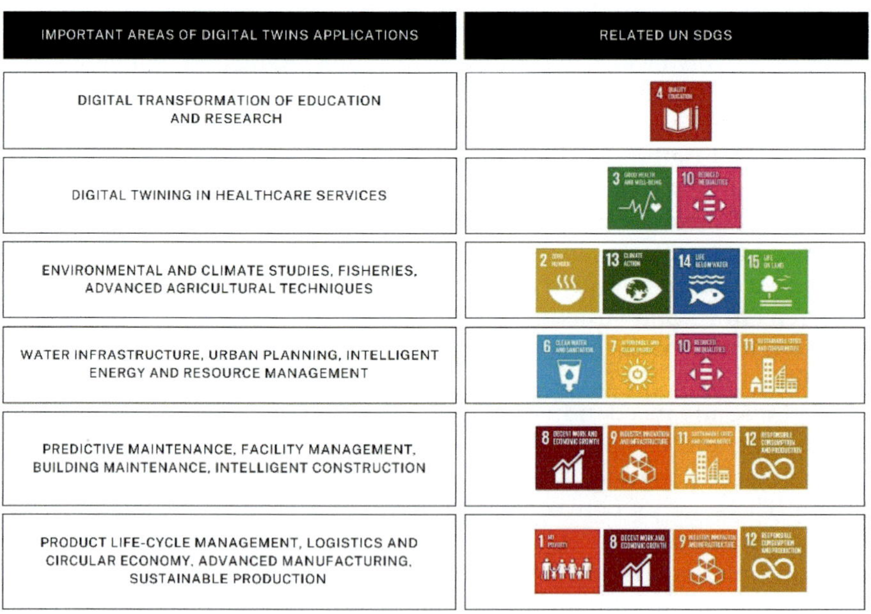

Figure 3: Important areas of Digital Twin applications and their related UN SDGS.

References

Al-Weshah, R. A., Saidan, M. N. and Al-Omari, A. S. (2016). Environmental ethics as a tool for sustainable water resource management. *Journal—American Water Works Association*, 108(3): E175–E181. https://doi.org/10.5942/JAWWA.2016.108.0037.

Alamelu, K. (2013). Digital inclusion—A conceptual framework. *International Journal of Advanced Research in Management and Social Sciences*, 2(12): 228–248.

Ali, Z., Saad, S., Rasheed, K. and Ammad, S. (2024). AI future perspectives and trends in construction. In *AI in Material Science* (pp. 239–261). CRC Press. https://doi.org/10.1201/9781003438489-11.

Allam, Z. and Jones, D. S. (2021). Future (post-COVID) digital, smart and sustainable cities in the wake of 6G: Digital twins, immersive realities and new urban economies. *Land Use Policy*, 101: 105201. https://doi.org/10.1016/j.landusepol.2020.105201.

Ammad, S., Saad, S., Bashir, M. T., Qureshi, A. H., Altaf, M. et al. (2021). Construction accidents via integrating Building Information Modelling (BIM) with emerging digital technologies a review. *2021 3rd International Sustainability and Resilience Conference: Climate Change*. https://doi.org/10.1109/IEEECONF53624.2021.9668066.

Arzo, S. T., Sikeridis, D., Devetsikiotis, M., Granelli, F., Fierro, R. et al. (2023). Essential technologies and concepts for massive space exploration: Challenges and opportunities. *IEEE Transactions on Aerospace and Electronic Systems*, 59(1): 3–29. https://doi.org/10.1109/TAES.2022.3169126.

Chakraborty, M., Kettle, J. and Dahiya, R. (2022). Electronic waste reduction through devices and printed circuit boards designed for circularity. *IEEE Journal on Flexible Electronics*, 1(1): 4–23. https://doi.org/10.1109/JFLEX.2022.3159258.

De Kerckhove, D. (2021). The personal digital twin, ethical considerations. *Philosophical Transactions of the Royal Society A*, 379(2207). https://doi.org/10.1098/RSTA.2020.0367.

Domenech, T. and Bahn-Walkowiak, B. (2019). Transition towards a resource efficient circular economy in Europe: Policy lessons from the eu and the member states. *Ecological Economics*, 155: 7–19. https://doi.org/10.1016/J.ECOLECON.2017.11.001.

Filgueiras, F. (2015). *Academic Paper Transparency and Accountability: Principles and Rules for the Construction of Publicity*. https://doi.org/10.1002/pa.

Filgueiras, F. (2016). Transparency and accountability: Principles and rules for the construction of publicity. *Journal of Public Affairs*, 16(2): 192–202. https://doi.org/10.1002/PA.1575.

Haris, M., Saad, S., Ammad, S. and Rasheed, K. (2024). AI in fabrication and construction. In *AI in Material Science* (pp. 169–192). CRC Press. https://doi.org/10.1201/9781003438489-8.

Haris, M., Saad, S., Ammad, S., Rasheed, K., Aman, H. et al. (2024). Integrating sustainable practices for effective solid waste management in peshawar: Overcoming challenges and promoting environmental. *2024 ASU International Conference in Emerging Technologies for Sustainability and Intelligent Systems (ICETSIS)*, 1–7. https://doi.org/10.1109/ICETSIS61505.2024.10459445.

Heimstädt, M. and Dobusch, L. (2020). Transparency and accountability: Causal, critical and constructive perspectives. *Https://Doi.Org/10.1177/2631787720964216*, 1(4): 263178772096421. https://doi.org/10.1177/2631787720964216.

Khan, M. I., Khan, N., Saad, S. and Khan, M. W. (2024). Artificial Intelligence in construction materials. In *AI in Material Science* (pp. 47–64). CRC Press. https://doi.org/10.1201/9781003438489-3.

Kim, Y. J., Kim, K. and Lee, S. K. (2017). The rise of technological unemployment and its implications on the future macroeconomic landscape. *Futures*, 87: 1–9. https://doi.org/10.1016/J.FUTURES.2017.01.003.

Li, W., Chien, F., Hsu, C. C., Zhang, Y. Q., Nawaz, M. A. et al. (2021). Nexus between energy poverty and energy efficiency: Estimating the long-run dynamics. *Resources Policy*, 72: 102063. https://doi.org/10.1016/J.RESOURPOL.2021.102063.

LNC Prakash, K., Ravva, S. K., Rathnamma, M. V. and Suryanarayana, G. (2023). AI applications of drones. In *Drone Technology* (pp. 153–182). Wiley. https://doi.org/10.1002/9781394168002.ch7.

Manifold, J., Renukappa, S., Suresh, S. and Georgakis, P. (2022). *How the UK Transportation Sector can Achieve Net Carbon Zero using Building Information Modelling*. British Academy of Management. https://wlv.openrepository.com/bitstream/2436/624816/3/BAM_2022_Joel et al_Final.pdf.

Naeem, U. H., Bilal, M., Syed, F., Rasheed, K., Saad, S. et al. (2022). A multilayer encryption model to protect healthcare data in cloud environment. *2022 International Conference on Data Analytics for Business and Industry, ICDABI 2022*. https://doi.org/10.1109/ICDABI56818.2022.10041708.

Nasibulina, A. (2015). Education for sustainable development and environmental ethics. *Procedia—Social and Behavioral Sciences*, 214: 1077–1082. https://doi.org/10.1016/J.SBSPRO.2015.11.708.

Omotayo, T., Egbelakin, T., Ogunmakinde, O. and Sojobi, A. (2024). *Innovations, Disruptions and Future Trends in the Global Construction Industry*. Routledge. https://doi.org/10.1201/9781003372233.

Pater, J. and Stadnicka, D. (2021). Towards digital twins development and implementation to support sustainability—Systematic literature review. *Management and Production Engineering Review*, 12(3): 63–73. https://doi.org/10.24425/mper.2021.138531.

Politou, E., Alepis, E., Virvou, M. and Patsakis, C. (2022). *Privacy and Data Protection Challenges in the Distributed Era. 26*. https://doi.org/10.1007/978-3-030-85443-0.

Rasheed, K., Saad, S., Ammad, S. and Bashir, M. T. (2024). Industry 4.0 and construction. In *AI in Material Science* (pp. 65–87). CRC Press. https://doi.org/10.1201/9781003438489-4.

Rasheed, K., Waqas, M., Saad, S., Zaland, A., Ali, Z. et al. (2024). Ethical and societal implications of AI in construction. In *AI in Material Science* (pp. 216–238). CRC Press. https://doi.org/10.1201/9781003438489-10.

Rasheed, K., Zaland, A., Saad, S., Ammad, S., Rostami, A. et al. (2024). History of AI. In *AI in Material Science* (pp. 15–46). CRC Press. https://doi.org/10.1201/9781003438489-2.

Saad, S., Haris, M., Ammad, S. and Rasheed, K. (2024). AI-assisted building design. In *AI in Material Science* (pp. 143–168). CRC Press. https://doi.org/10.1201/9781003438489-7.

Saad, S., Rasheed, K., Ammad, S., Ali, Z., Zaland, A. et al. (2024). The conclusive role of AI in Industry 4.0 construction materials. In *AI in Material Science* (pp. 262–270). CRC Press. https://doi.org/10.1201/9781003438489-12.

Saad, S., Rasheed, K., Ammad, S., Waqas Khan, M., Zaland, A. et al. (2024). Is AI the architect of tomorrow's materials in the age of Industry 4.0? In *AI in Material Science* (pp. 1–14). CRC Press. https://doi.org/10.1201/9781003438489-1.

Saad, S., Zaland, A., Ammad, S. and Rasheed, K. (2024). Robotics in construction. In *AI in Material Science* (pp. 115–142). CRC Press. https://doi.org/10.1201/9781003438489-6.

Singh, S. K., Chen, J., Del Giudice, M. and El-Kassar, A. N. (2019). Environmental ethics, environmental performance, and competitive advantage: Role of environmental training. *Technological Forecasting and Social Change*, 146: 203–211. https://doi.org/10.1016/J.TECHFORE.2019.05.032.

Smith, C. (2015). An analysis of digital inclusion projects: Three crucial factors and four key components. *Journal of Information Technology Education: Research*, 14(2015): 179–188. https://doi.org/10.28945/2262.

Waltersmann, L., Kiemel, S., Stuhlsatz, J., Sauer, A., Miehe, R. et al. (2021). Artificial Intelligence applications for increasing resource efficiency in manufacturing companies—A comprehensive review. *Sustainability*, 13(12): 6689. https://doi.org/10.3390/SU13126689.

Waqas Khan, M., Saad, S., Ammad, S., Rasheed, K., Jamal, Q. et al. (2024). Smart infrastructure and AI. In *AI in Material Science* (pp. 193–215). CRC Press. https://doi.org/10.1201/9781003438489-9.

Zaland, A., Saad, S., Ahmad, W., Ammad, S., Khan, M. W. et al. (2024). Enhancing mechanical properties of concrete by adding polypropylene fibers. *2024 ASU International Conference in Emerging Technologies for Sustainability and Intelligent Systems (ICETSIS)*, 771–778. https://doi.org/10.1109/ICETSIS61505.2024.10459663.

Zaland, A., Saad, S., Rasheed, K. and Ammad, S. (2024). Automation in construction industry. In *AI in Material Science* (pp. 88–114). CRC Press. https://doi.org/10.1201/9781003438489-5.

Chapter 10

Shaping the Future of Construction with Advanced Technologies

Muhammad Haris,[1] *Syed Saad*,[2] *Kumeel Rasheed*,[1]
Syed Ammad[3,*] and *Muhammad Waqas Khan*[1]

Introduction

The construction industry is undergoing a transformative phase driven by advanced technologies that are reshaping how projects are designed, managed, and executed. Innovations like Building Information Modeling (BIM), 3D printing, drones, and Artificial Intelligence (AI) are revolutionizing construction practices, improving efficiency, reducing costs, and enhancing safety. For instance, BIM allows for the creation of detailed 3D models that facilitate better project visualization and collaboration among stakeholders. In the UK, the Crossrail project utilized BIM to manage complex data, leading to significant time and cost savings. Similarly, 3D printing is being used to construct buildings more quickly and with reduced waste. The Office of the Future in Dubai, a fully functional office building, was constructed using 3D printing

[1] Department of Civil Engineering, CECOS University of IT and Emerging Sciences, Pakistan.
[2] Science, Engineering & Manufacturing Centre Sandwell College West Bromwich, United Kingdom.
[3] School of Engineering and Built Environment, Griffith University, Gold Coast, Australia.
* Corresponding author: syed.ammad@griffithuni.edu.au

technology in just 17 days. Drones are also making a significant impact by conducting site surveys and inspections more efficiently and safely. For example, drones were used in the construction of the Sacramento Kings' Golden 1 Center, allowing for precise and timely site inspections. AI and machine learning algorithms are being employed to predict project risks, optimize resource allocation, and enhance decision-making (Saad et al., 2024). One notable example is the use of AI by the company Vinci Construction to predict and mitigate potential project delays and cost overruns. These real-life case studies underscore the potential of advanced technologies to shape the future of construction, making it more sustainable, efficient, and innovative.

The discussion on the future of construction within the realm of advanced technologies highlights critical perspectives on industry evolution, particularly focusing on the link between information technology (IT) and the construction field. Researchers and industry experts have made significant efforts in understanding how new technologies are set to transform and guide the construction industry. Past research mainly explored direct causal relationships between emerging technologies and the sector's future direction. It focused on predicting future construction scenarios based on cutting-edge IT tools, considering specific technologies' potential to shape future construction processes. These studies imagined concepts like the nD-enabled construction idea (Ding et al., 2014) and proposed modeling roadmaps for advancements within the 2004–2012 timeframe. There was a combined effort to create visions of IT's role in construction based on extensive literature review, expert consultations, and industry feedback.

Contrary to previous methods, current research stands out based on two key principles guiding its path. Firstly, it asserts that grasping the future of construction IT necessitates an extensive examination of potential global scenarios and their implications for the construction industry and information technology (Haris et al., 2024). Understanding the forces driving change involves a global exploration, narrowing down to focus on the construction industry and specifically targeting construction IT. Secondly, it argues that construction IT's direction is not solely determined by IT advancements but is significantly influenced by broader technological developments and socio-economic, environmental, and political factors. Therefore, a modified La Prospective scenario-planning approach was used, combining strategic discussions, surveys, prospective workshops, and horizon scanning to analyze and understand these driving forces (Erdogan et al., 2010).

Moreover, several studies take a forward-looking perspective by examining long-term planning, using scenario planning to imagine possible futures up to 25 years ahead. The selection of 2030 as the target horizon is due to its practicality within the scenario planning framework, providing a balance between the need for predictability and the limits of extending current trends. Several research structures involve an analysis of the key drivers, major trends, and factors influencing the future by considering social, technological, environmental, economic, and political aspects (Rasheed et al., 2023). Four future scenarios are carefully created, each starting with a global view of 2030 and then exploring how these global predictions affect the construction industry and, in turn, how ICT will be involved in construction (Adel, 2023).

Eight themes identified as in (Erdogan et al., 2010) are deemed critical for shaping the future. Three themes, specifically global environmental change, future financial frameworks, and seismic power shifts, are highlighted as crucial. These themes serve as a foundation for scenario logics, directing the creation of future scenarios. The future of construction IT is expected to enable less geographically dependent work practices, enhance collaboration through technologies like virtual reality (VR) and holograms, and offer a better work-life balance through wireless and intelligent frameworks. The industry is foreseen to shift its focus from manual labor to innovation and design supported by IT, possibly incorporating artificial intelligence to augment human creativity. Anticipated changes include increased use of robots in construction, off-site construction practices, and comprehensive sensory simulations for design. Intelligent buildings managing themselves through embedded sensors and proactive maintenance are expected to be mainstream. Project management, learning, and adaptation through games and role plays are anticipated, integrating IT into daily life effortlessly for end-users (Raja Santhi and Muthuswamy, 2023).

The key areas for action, such as managing change and fostering innovation through improved communication, education, process enhancement, interoperable IT systems, user-centered IT, and sustainability are necessary and the need for better education to support a shift from a craft-based industry to a knowledge-intensive model (Bilal et al., 2023) and advocates for integrated solutions to address process inefficiencies in the industry cannot be ignored.

Finally, the scenario-planning methodology is shown to be a critical strategy for developing a comprehensive vision for the development of IT in the construction industry. This paradigm emphasizes the necessity of ongoing evaluation and adaptive improvement, and it calls for a dynamic conceptualization of future trajectories within the construction industry.

Adapting to the constantly changing world of industrial dynamics and technological breakthroughs requires an iterative process like this one. The construction industry's inherent volatility requires a model that acknowledges the rapid pace of technological change and its impact on industry needs. This is exemplified by the scenario-planning technique, which encourages a detailed analysis of potential future scenarios, considering various technological advancements and socioeconomic factors. This method enables a deep understanding of the possible paths the construction IT landscape may take, allowing stakeholders to anticipate and adapt to different situations.

The procedure also highlights how important it is to create and maintain a strategic plan. For the construction industry, this roadmap acts as a compass, pointing the way through the complicated interactions between market demands, technology advancements, and regulatory frameworks. A roadmap of this kind helps to match industry practices and research projects with the future by outlining the major milestones, objectives, and trajectories (Raja Santhi and Muthuswamy, 2023). Moreover, the scenario-based strategic roadmap serves as a driving force behind well-informed decision-making in the field of construction information technology. It provides stakeholders with a formal framework for assessing and ranking research directions, investment plans, and technology interventions. Through the explanation of likely obstacles, chances, and possible disturbances, this roadmap enables participants to actively negotiate the changing environment, promoting flexibility and resilience in the sector.

The iterative process of scenario planning emphasizes the value of flexibility and encourages frequent review. Because the construction IT ecosystem is naturally vulnerable to sudden changes and unanticipated events, it requires agility to adapt to new trends and problems. As a result, this strategy suggests an ongoing cycle of study, reassessment, and improvement to make sure the future that is envisaged is adaptable and considers changing technology environments and industrial demands. Essentially, scenario-planning emphasizes the importance of imagining the future broadly and planning strategically for construction IT. It encourages adaptability, smart decision-making, and planning ahead, aiming to make the construction industry more flexible and readier to use new technologies for sustainable progress.

Emerging Technologies in Digital Twins and Robotics

The historical context of DTs dates back to the 1960s in space robotics, with National Aeronautics and Space Administration (NASA) utilizing early forms known as "living models." The term "digital twin" was coined in 2002, evolving into a concept to improve product lifecycle management (PLM). The surge of interest in DTs since 2003, driven by Industry 4.0 initiatives and technological advancements like IoT and big data, indicates their growing significance. The market value for digital twins was projected to grow significantly, indicating their relevance in various sectors (Mazumder et al., 2023).

DTs offer unprecedented potential due to their persistence, insight, and explain-ability (explainability), allowing for continuous data flow between physical systems and their digital replicas. They enable predictive analysis, troubleshooting, and a deeper understanding of system interactions, contributing significantly to industrial safety, design enhancement, and future simulations. Moreover, recent developments highlight the shift from closed systems to interconnected modular frameworks, enabling broader data access and enhancing process understanding (Khan et al., 2024). The evolution of DT technology and its integration into various domains of robotics may begin by tracing the development of DT from its inception in space robotics to its contemporary applications in real-time robotic manufacturing ecosystems. The five prominent domains where DT has seen significant implementation: space robotics, medical and rehabilitation robotics, soft robotics, human-robot interaction, and industrial robotics (Huang et al., 2021).

In the domain of space robotics, the various implementations of DT, such as virtual testbeds (VTBs) for optical sensors, haptic telerobotic assembly of modular satellites, spacecraft maintenance, repair, and operations (MRO), and the autonomy of robotic spacecraft for deep space exploration, including NASA's Artemis program (Mazumder et al., 2023). Moving to medical and rehabilitation robotics, the growing application of DT in medical telerobotic, VR-assisted surgeries, robotic rehabilitation, and AI implementation for predictive analysis and system control are noticeable. Notably, the COVID-19 pandemic accelerated developments in medical telerobotic and robotic assembly of medical equipment (Huang et al., 2021).

In the realm soft robotics, the application of digital twins (DT) in virtual reality (VR), the reconstruction of soft robot digital twins (DT) in VR environments, AI methodologies, the use of Triboelectric Nanogenerator (TENG) sensors, and multi-modal interaction is of great significance. It emphasizes the importance of seamless data transmission between physical robots and their virtual twins, especially in adverse conditions (Mazumder et al., 2023). Across these domains, digital twins have propelled robotics forward by facilitating immersive VR applications, predictive AI models, advanced data transmission through innovative sensors, and multi-modal interaction techniques. It showcases the potential of DT to revolutionize various aspects of robotics, from surgical procedures to space exploration, by providing a detailed insight into recent trends and technological advancements (Gallala et al., 2022).

Challenges and Limitations

Some of the identified constraints address domain-specific challenges, such as latency issues, complexity in soft robotics, stringent precision requirements in medical robotics, and computational demands in organizational integration. This comprehensively examines the expanding landscape of DT in conjunction with robotics, outlining evolving technologies, applications across diverse sectors, and the pertinent challenges shaping these domains (Mazumder et al., 2023). The exploration of emerging technologies can be observed in Table 1.

The emerging trends and potential research scopes in the integration of DT with robotics highlights various areas where advancements are rapidly occurring, including immersive VR, haptic telerobotic, DT-aided AI implementations, and robot work-cell simulation. These technologies are noted for their potential in enhancing workspace visualization,

Table 1: Emerging applications of Digital Twins in robotics domains.

Domain	Implementations and Applications	Key Benefits
Space Robotics	Virtual testbeds for optical sensors, haptic telerobotic assembly of modular satellites, spacecraft MRO, autonomy in deep space exploration (e.g., NASA's Artemis program).	Enhances spacecraft operations, supports deep space exploration, improves assembly efficiency.
Medical & Rehabilitation	Medical tele robotics, VR-assisted surgeries, robotic rehabilitation, AI for predictive analysis and system control, accelerated developments during the COVID-19 pandemic.	Improves surgical precision, enhances rehabilitation outcomes, supports remote medical procedures.
Soft Robotics	VR applications, reconstruction of soft robot DTs in VR, AI integration, Triboelectric Nanogenerator sensors, multi-modal interaction.	Enables adaptive robots, enhances data transmission in adverse conditions, supports VR integration.
Human-Robot Interaction	Enhances immersive user experiences, improves AI models for predictive analysis, supports multi-modal interaction techniques.	Facilitates intuitive human-robot collaboration, improves interaction efficiency.
Industrial Robotics	Integration into real-time manufacturing ecosystems, predictive maintenance, system optimization, data-driven process improvements.	Optimizes production efficiency, reduces downtime, enhances operational insights.

enabling remote control of robots, and improving robot maintenance and performance prediction (Mazumder et al., 2023).

Moreover, the challenges associated with these advancements, such as the complexity of constructing Rough set based Decision Tree (RDT) models, security concerns, computational intensity, and network connectivity issues. Solutions to these challenges are proposed, including using blockchain for security, developing less computationally intensive frameworks, and enhancing network connectivity. Additionally, the future of DT-incorporated robotics lies in unified frameworks that combine emerging practices. The integration of technologies like blockchain and Metaverse is proposed as a potential destination for future robotics, especially in creating next-generation VTBs. This approach aims to integrate immersive experiences with AI, haptic control, and VR applications to optimize robotic systems. The importance of these emerging trends and proposed frameworks in shaping the future of robotics. The limitations of the current review proposes further research to encompass omitted areas in robotics, intending to have a significant impact on future research in DT-incorporated robotics.

Global Perspectives and Industry Trends

The significance of Industry 4.0 and smart manufacturing in the context of the global economy is very crucial. It emphasizes the aim of Industry 4.0 to establish a networked architecture addressing compatibility issues in automation systems, enhancing flexibility in manufacturing. This framework aligns with smart manufacturing, where advanced robotics plays a vital role (Huang et al., 2021). In recent times, the need for sustainable development due to environmental challenges like carbon emissions has necessitated a shift from traditional economic growth. Achieving sustainability involves balancing financial, environmental, social, and governance dimensions (FESG factors), posing challenges for manufacturing enterprises. In response, AI-powered DT technology is seen as a solution, adapting model-based approaches to support decision-making in evolving scenarios (Ghosh and Ghosh, 2021).

The gaps in existing research, particularly the lack of an industry-focused review of "AI + DT" technologies concerning sustainability and circular economy aspects. It poses research questions (RQs) to address these gaps, aiming to explore the current state, benefits, challenges, and future prospects of AI-enabled DTs in smart manufacturing and advanced robotics. We can delve into definitions such as digital twin, digital shadow, and digital thread, providing insights into their origins and functionalities. Focusing on AI-driven DT technologies in Industry 4.0 does not cover standalone IoT or communication technologies like 5G.

The evolution of manufacturing industries, emphasizing the shift towards sustainable resilient manufacturing practices within the context of Industry 4.0 faces the challenges faced by manufacturing industries in balancing objectives like productivity, availability, and quality amidst ever-changing market conditions (Waqas Khan et al., 2024). There's a growing focus on digital production twins, enabled by simulation models and pervasive connectivity, as a crucial element in future manufacturing transformations driven by big data (Ghosh and Ghosh, 2021). Within the framework of circular economy and sustainability goals (e.g., European Green Deal), the emphasis is on the concept of "sustainable productivity," incorporating FESG factors for assessing sustainable production by the various aspects of DTs and their integration with AI across different manufacturing levels: factory floor, machinery/equipment, and processes/materials (Huang et al., 2021). At the factory floor level, discussions revolve around the use of automated production systems, AI, sensors, and cloud computing to enable modularization, reconfiguration, and optimization of production environments. AI integration within DTs aids in production planning, control, and quality management, showcasing methods like decision trees, deep Q-networks, reinforcement learning, and neural networks to enhance processes and resource allocation.

The machinery/equipment delves into condition monitoring, predictive maintenance, and machine dynamics while AI integration in this context focuses on the use of machine learning (ML) and deep learning (DL) methods to identify fault conditions, estimate remaining useful life, and optimize control systems, aiding in cost-effective predictive maintenance strategies. Regarding processes/materials, the AI-driven digital process twins in various manufacturing techniques, including metal cutting, additive manufacturing, and composite material processing can be examined. These AI-integrated models aid in optimizing process design, controlling quality, and accelerating production ramp-up by learning from heterogeneous data sources and leveraging ML/DL for real-time simulations (Huang et al., 2021).

The need for standard frameworks for multi-scale DTs, data interoperability, high-fidelity lightweight models, and smart services/business models is very crucial. It emphasizes the necessity of interdisciplinary cooperation, data contextualization, and the development of data-driven approaches to address uncertainties and extend model boundaries, ultimately promoting sustainable, resilient manufacturing (Ghosh and Ghosh, 2021).

The current trends and perspectives in the global robotics industry can delve into the increasing significance of DT for robots in various applications, emphasizing their crucial role in multi-robot coordination, safe human-robot interaction (HRI), and complex human-robot collaboration (HRC). The adoption of traditional simulation frameworks and cloud-based approaches for robot DT implementations is highlighted, showcasing applications in kinematics, communication, control, planning, energy modeling, and various robot use cases like welding, cleaning, assembly, manufacturing, maintenance, and construction (Huang et al., 2021).

AI-equipped DTs are gaining traction, particularly in enabling semi- and fully autonomous robotic systems through concepts like transfer learning and imitation learning (Rasheed et al., 2024). These data-driven and AI-equipped DTs are proving invaluable, especially for complex robotic systems where building high-fidelity dynamics models is unfeasible. They have applications not only in traditional robotics but also in biomimetic robotic system development, such as in the case of robotic fish.

AI-equipped DTs based on learning algorithms and subfields like control, planning, HRI, and HRC can be explored by how feedback and accurate information collected from physical sensors aid in robotic control, highlighting instances where AI-driven approaches enhance sensing capabilities and improve system understanding for better task performance. It also delves into the significance of high-level robotic planning and how reinforcement learning integrated with DTs is transforming complex systems' planning. Furthermore, the importance of DTs in ensuring safer human-robot interaction and collaboration in Industry 4.0 contexts and how AI-equipped DTs excel in responding to uncertainties and diverse human behaviors, thus enhancing safety and efficiency in collaborative scenarios is apparent. Applications range from VR-based collaborative construction work to human standing-posture recognition in HRC (Huang et al., 2021). Additionally, the robot maintenance and other applications, showcasing how data-driven methods, deep learning, and predictive maintenance strategies are being integrated into robotic systems for anomaly detection, health monitoring, and customized maintenance can't be ignored either.

Research and Development Opportunities

The significant research and development opportunities within the construction sector by leveraging advanced technologies include several key areas of focus like:

Technological Applications

The modern construction environment demonstrates a notable propensity to leverage emerging technology in order to reshape and enhance its operating frameworks. Research conducted on this subject very much supports the need for implementing and incorporating multiple advanced technologies. With blockchain, AI, BIM, sensors and automation as the front runners of innovation in the construction industry is stepping into a new age. These are intentional adoptions of technology which have the potential to revolutionize this industry by ushering in previously unheard-of levels of productivity , accuracy and efficiency.

With its reputation for being decentralized and unchangeable, blockchain technology has become a potentially useful instrument in the building industry. Its use covers a wide range of areas, including supply chain management, smart contracts, and safe and transparent transactions. The construction industry wants to use blockchain to reduce risks related to supply chain operations and contractual agreements, improve trust, and expedite processes (Kousi et al., 2019). Simultaneously, the integration of artificial intelligence presents a transformative paradigm for the sector. AI algorithms exhibit the capacity to analyze vast datasets, predict patterns, and optimize decision-making processes. In construction, AI-powered systems streamline project management, automate repetitive tasks, and offer predictive maintenance solutions (Saad et al., 2024). Also, the use of AI encourages risk analysis, allocation of resources and smart scheduling that all contribute to enhancing project outcomes as well as operation effectiveness (Babaeian Jelodar and Sutrisna, 2022).

One of the main technological innovations that are transforming how construction is performed is building information modeling (BIM). BIM transforms actual infrastructure into comprehensive digital representations, hence revolutionizing project planning and execution. Through the use of this technology, stakeholders may discuss, simulate, and visualize designs, improving communication and coordination all the way through the project lifecycle. By combining various aspects of construction, BIM reduces errors, increases decision-making capabilities, and saves money (Babaeian Jelodar and Sutrisna, 2022). Sensors, a part of IoT applications, are an important feature when we talk about construction technology. These sensors, which are placed in the building sites, machines and structures enable real-time data collection. They offer information about structural integrity as well as environmental conditions and equipment performance. By employing sensors in construction, safety measures are enhanced and predictive maintenance and data-based decision making is facilitated (Kousi et al., 2019).

Additionally, automation within construction processes serves as a catalyst for efficiency enhancement. Automated machinery, robotics, and autonomous systems reduce manual labor, enhance precision, and expedite tasks. From assembly line operations to complex construction activities, automation optimizes workflows, minimizes errors, and accelerates project timelines (Babaeian Jelodar and Sutrisna, 2022). These advanced technologies within construction activities signify an important shift toward a more sophisticated and efficient industry. Their collective integration and utilization herald a future where precision, efficiency, and innovation converge to reshape the construction landscape, driving unprecedented progress and excellence within the sector.

Digital Transformation in Construction

Digital transformation in the context of the construction sector aims at incorporating digital solutions to revolutionize methods of implementing construction projects. This change initiative is meant to address the key challenges that involve gains in productivity, communication, and decision making all through the life cycle of the project. For example, many firms adopted elements such as BIM which improve project visualization, decrease mistakes, and optimize construction procedures. Another example that applies this is the use of drones for site surveying and monitoring which enhances precision and safety, cuts time and costs. Apart from the efficiency aspect, it facilitates analysis of results in real time, as well as predictive maintenance which results in progressive improvements in projects and greater client satisfaction.

At the core of this transformation lies a deliberate drive to curtail human involvement in operational workflows within the construction domain. Embracing automation, robotics, and AI-driven systems, the industry is navigating towards a landscape where manual tasks are progressively supplanted by innovative technological solutions. This concerted effort to reduce human intervention not only streamlines operations but also enhances precision, productivity, and safety across construction sites (Babaeian Jelodar and Sutrisna, 2022). An important aspect of this shift is the enhancement of Industry 4.0 into the construction sector. This new industrial model integrates digital technology with the conventional forms of manufacturing and industries, resulting in a system integration that is driven by data. The construction industry's goals of IoT, cloud computing and cognitive computing is to establish a seamless interconnected network with real-time intelligent data exchange and data usage for unmatched productivity and creativity.

At the same time, embedding IoT tools into construction processes becomes a key element of this digital transformation. IoT devices and sensors within construction sites, machinery, structures can be utilized to capture and relay real-time data (Saad et al., 2024). These devices serve as the keystone to establish a network of physical assets that is interconnected, providing data-informed understandings on operational parameters, structural health and environmental conditions. By using IoT tools, construction stakeholders gain invaluable information that allows them to make informed predictive maintenance decisions and proactive risk mitigation strategies (Rasheed et al., 2024). Furthermore, the main goal of the digital revolution in construction is to decrease and, when possible, eliminate errors found in traditional methods. By integrating technology into all aspects of construction, the industry aims to remove inefficiencies, reduce rework, and improve the overall quality of projects. Using digital strategies and smart systems promises to create an environment where precision, accuracy, and reliability are built-in qualities, ensuring projects are executed smoothly with fewer mistakes (Vial, 2021).

In essence, this transformative journey toward digital strategies and smart systems signifies an important leap forward for the construction industry. It represents an era where technology not only augments but fundamentally reshapes traditional practices, fostering a future characterized by unparalleled efficiency, precision, and sustainable growth within the construction landscape.

Infrastructure Development

Infrastructure development involves physical works hence involves the planning, establishing and sustenance of structures and systems in support of societies and economies. These are transportation infrastructure such as roads, bridges and railways, infrastructural utilities such as water and sewerage management, communication infrastructure and facilities including telecommunications and internet connectivity and social infrastructure in the form of schools and hospitals among others. Infrastructure needs to be well constructed for economic and social benefits as it provides access to trade, strengthens connectivity and increases the standards of living among the populace. For example, the construction of high-speed rail like those of Japan and China to enable traveling between major cities has cut down on the time, hence impacting the economy, productivity and tourism. Some cities like Barcelona in Spain have smart infrastructure technology for traffic management, waste disposal and energy efficiency improvements thus improving their livability. Construction of hydropower dams like the Three Gorges Dam upstream China, has offered renewable energy and prevented flooding with irrigation and water supply.

The sector might move towards resilience and navigate these challenges using new technical approaches. These methods offer hope after major disruptions. Advanced technology applications can effectively minimize disruptions, enhance productivity, and address infrastructure investment gaps (Thacker et al., 2019). These technologies include BIM evolution and improvement, sensing technologies, automation, blockchain, and AI, which are improving the construction industry's capacity to manage vulnerabilities. Again, due to the unique characteristics of the blockchain, namely decentralization of records and high levels of protection, transactions remain transparent and irreversible and increase the ability to manage project financing in an environment of high market uncertainty and risk. On the other hand, the implementation of AI provides the ability of predictive analysis, efficient organizational flow, and effective management of resources which in turn decreases the industry's vulnerability to financial volatility (Babaeian Jelodar and Sutrisna, 2022).

Furthermore, by incorporating BIM frameworks, sensors, and automation into construction activities, a range of opportunities is revealed. This includes improving digital accuracy, gaining real-time data insights, and increasing operational efficiencies. These technologies help to navigate challenges such as pandemics or financial crises. They enable remote project management, safer on-site operations, and informed decision-making, guiding the sector through uncertain times.

Global Spending and Innovation

The acknowledgment of substantial global infrastructure spending, along with the evident gap in meeting future spending demands, highlights the urgent need for increased innovation, advanced work methods, and enhanced collaboration. This recognition not only outlines current financial commitments but also stresses the necessity for proactive measures to bridge the gap between current spending capabilities and growing future needs (Babaeian Jelodar and Sutrisna, 2022). This illustrates innovative approaches and transformative strategies in construction and infrastructure development. It emphasizes the need for more efficient methods, embracing advanced technologies, and fostering collaboration to reduce spending gaps. By addressing the substantial investments needed and current disparities, the focus shifts to strategic advancements that promote sustainability and global infrastructure growth through innovation and collaboration (Ali et al., 2024).

To close the spending gap, companies, governments, and numerous stakeholders must work together to optimize resource allocation, streamline procedures, and implement forward-thinking technological solutions. It is not solely about acknowledging the financial magnitude but also recognizing the need for groundbreaking methodologies that can significantly augment the efficiency and effectiveness of infrastructure development. Furthermore, this awareness implies a change toward embracing creative frameworks, enabling technology, and collaborative activities, prompting a reevaluation of traditional techniques. It emphasizes the importance of integrated strategies that use emerging technology, build cross-sector relationships, and harness the potential of innovation to generate long-term, scalable solutions.

Among the significant spending gaps and growing infrastructure needs, innovation is essential for meeting these challenges. It is not just about recognizing the spending disparity but actively fostering an environment of innovation and collaboration. This can effectively transform and revolutionize global infrastructure development to meet future demands (Babaeian Jelodar and Sutrisna, 2022).

Design Thinking in Digital Twins and Robotics

Design thinking in digital twins and robotics involves several crucial phases and considerations essential for constructing and implementing a DT system as shown in Figure 1.

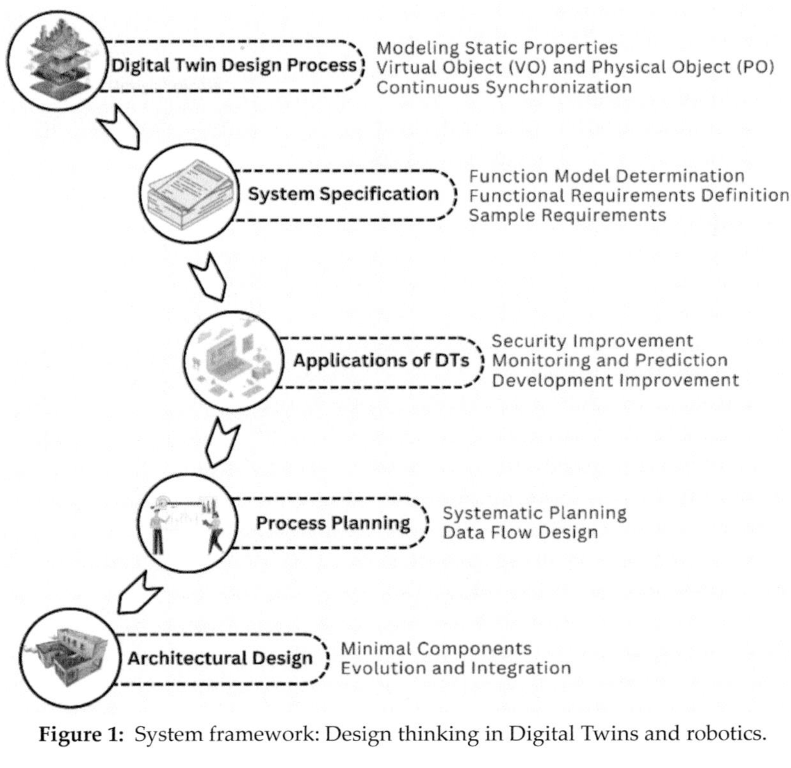

Figure 1: System framework: Design thinking in Digital Twins and robotics.

Digital Twin Design Process

Modeling static properties involves defining system requirements, functions, and constraints by analyzing data flows, system architecture, technical needs (hardware/software), and communication aspects (Segovia and Garcia-Alfaro, 2022). The Virtual Object (VO) in a DT mirrors the Physical Object (PO) throughout its life cycle. If the PO exists during design, early efforts focus on linking the VO to the PO. If the PO doesn't exist, a prototype PO helps guide the VO's creation, ultimately ensuring the VO mirrors the PO or its prototype. DTs also maintain continuous synchronization with the production system, requiring flexibility during system design to handle updates or changes in engineering models and processes (Rasheed et al., 2024).

System Specification

This phase outlines the DT's functional objectives and requirements, such as optimization, security, monitoring, or prediction of physical processes. It involves determining included functions, modeling aspects, and data exchange between components (Segovia and Garcia-Alfaro, 2022). Accurately defining DT requirements based on the PO and extracting necessary information is crucial. Examples include optimizing processes, enhancing security, and monitoring physical systems. This includes various requirements for DT specifications, including optimization, security enhancement, and monitoring, citing research proposals for each scenario (Segovia and Garcia-Alfaro, 2022).

Applications of DTs

DTs optimize systems such as product distribution, network management, energy efficiency, emissions reduction, and production layout planning. They improve security against cyber-attacks by proposing resilient networks, attack detection strategies, and incident prevention mechanisms. DTs enable real-time monitoring, fault detection, system health analysis, and predictive maintenance across various sectors like manufacturing, infrastructure, and energy. They enhance product design, experimentation, process sustainability, and quality control. Additionally, DTs provide low-risk, low-cost environments for operator training, improving decision-making abilities and reducing operational errors (Segovia and Garcia-Alfaro, 2022).

Process Planning

Process planning involves detailing system activities, relationships between components, modeling system functions, and gathering data from physical processes, which is crucial for creating and maintaining digital twins (Segovia and Garcia-Alfaro, 2022). Additionally, designing data flow is vital for accurate digital twin modeling, which includes defining data sources, inputs, outputs, and communication patterns among system components (Segovia and Garcia-Alfaro, 2022).

Architectural Design

DTs usually include a PO, VO, and two-way data exchange. Different architectures suggest various components, such as control layers, data storage, security, services, integration, and cloud-based cyber layers (Segovia and Garcia-Alfaro, 2022). DT architectures must evolve to handle system changes, adapt behaviors, support multiple models, and integrate with other DTs or systems (Segovia and Garcia-Alfaro, 2022). The academic exploration of "Design Thinking in Digital Twins and Robotics" emphasizes thorough planning, comprehensive data handling, and adaptive architectures essential for DT implementation across diverse applications (Segovia and Garcia-Alfaro, 2022).

It is critical to apply design thinking in the context of digital twins and robots inside the NDE 4.0 (Non-Destructive Evaluation 4.0) paradigm. Design thinking is considered as a beneficial strategy for designing and implementing NDE 4.0 technologies. It is an iterative process intended at understanding consumers, challenging assumptions, and redefining challenges. Design thinking is employed in a nested manner and initially, it is used to craft a management approach that empathizes with the NDE community and outlines the strategy surrounding the "Why, What, and How" of NDE 4.0. Then, it is further incorporated into the adoption process to define a path from inspection insights to creative digitalization. The core philosophy of design thinking is visualized in the form of an iterative loop: Empathize, Define, Ideate, Prototype, and Test. This methodology enables a learning path starting from understanding end beneficiaries and culminating in viable solutions (Vrana and Singh, 2021).

Within the context of digital twins, the concept is detailed as a comprehensive repository encapsulating all pertinent data related to an asset or component. It not only stores data but enables simulation of the asset's usage and allows visualization of the stored data and simulation outcomes. Digital twins are categorized based on the assets they represent, such as production facilities, equipment, products, components, inspection systems, and more. Moreover, they can be layered and inheritable, providing a nuanced view of interconnected assets (Mazumder et al., 2023). Regarding robotics, the evolution from previous industrial revolutions to NDE 4.0 highlights the integration of robotics in enhanced automation within Industry 4.0, enabling machines to adapt to environments based on observations rather than predefined programming. Collaborative robots (cobots) exemplify this, allowing humans and robots to share a workspace without the need for physical barriers, supporting direct interaction (Vrana and Singh, 2021). Overall, the application of design thinking as a foundational strategy for developing NDE 4.0 technologies, particularly within the sector of digital twins and robotics, aiming to enhance the performance and innovation of these systems is of great importance.

Envisioning the Future of Digital Twins and Robotics in Smart Cities

It is the advanced usage of digital twins and robotics redefining the concept of smart cities and making urban life easier, sustainable, and efficient. Digital replicas are virtual representations of tangible items, business activities, or city infrastructure; they enable time-efficient, accurate modeling and forecasting of a city's operations and performance (Saad et al., 2024). For instance, Singapore has created a twin digital model of the whole country to boost urban planning and emergency management. As for robotics, city applications are being applied in services where human labor is deemed to be replaced for efficiency. Self-driving drones in Dubai as an example are deployed for security and monitoring of traffic situations and information is gathered specifically to reduce traffic jams and enhance safety on the roads. The described technologies help to apply the efficient and accurate maintenance of infrastructure, energy, and resources. In the emerging complex smart cities, it will be evident that future advanced applications of digital twins and robotics will be critical to solve urban issues, and to advance public service delivery and residents' quality of life. For example, Barcelona's smart city application that uses sensors and IoT in waste management has brought efficiency of cost and environmental returns.

Envisioning the future of digital twins and robotics in smart cities involves integrating advanced technologies for better urban management. Digital twins, virtual replicas of physical entities and activities, capture changes in urban environments using sensors, satellites, and other technologies (Wang et al., 2023). Robotics also plays a key role in smart cities, optimizing tasks and decision-making across various sectors like transportation, construction, and healthcare. Smart cities require resilient infrastructures using IoT devices and geographic information systems (GIS) for real-time data collection and analysis, aiding in urban planning and management (Zaland et al., 2024). As for the developments, upcoming emerging technologies for cyberGIS, BIM integration, laser scanning, machine learning, and high-performance computing will further improve these techniques (Rezaei et al., 2023).

Several issues like, analysis of extra-large datasets, integration with other solutions, and a shift to digital services have to be addressed to achieve this. The dream of scholars is to create simple tools and models for the practice of advanced sustainable development and the satisfaction of various urban demands (Wang et al., 2023). The historical process of cities' development, which satisfies people's needs for communication and interaction, economic development and technology illustrates the significance of digital revolutions. Smart cities were introduced in the mid-1990s with projects in Amsterdam and gained corporate attention from players like IBM and Cisco Systems in the early 2000s that aimed to improve the efficiency and quality of life in a city.

The potential of 6G technology for smart cities, offering significant advancements over 5G, is promising but requires matching infrastructure and technology (Allam and Jones, 2021). DTs and Robotics play a crucial role in smart cities, offering platforms to model and predict city changes. Current DTs often remain underutilized in decision-making, but the concept of Fused Twins (FTs), merging physical and digital twins through extended reality (XR), aims to make smart cities more accessible (Grübel et al., 2022). The evolution of smart cities, driven by ICT advancements, shows the need for DTs to simplify complex urban operations. Examples like smart waste disposal and parking systems illustrate urban technology progress (Javaid et al., 2023). Situated analytics and augmented reality integrate data visualization in physical spaces, enhancing user interaction and understanding.

References

Adel, A. (2023). Unlocking the future: Fostering human–machine collaboration and driving intelligent automation through Industry 5.0 in smart cities. *Smart Cities*, 6(5): 2742–2782. https://doi.org/10.3390/SMARTCITIES6050124.

Ali, Z., Saad, S., Rasheed, K. and Ammad, S. (2024). AI future perspectives and trends in construction. In *AI in Material Science* (pp. 239–261). CRC Press. https://doi.org/10.1201/9781003438489-11.

Allam, Z. and Jones, D. S. (2021). Future (post-COVID) digital, smart and sustainable cities in the wake of 6G: Digital twins, immersive realities and new urban economies. *Land Use Policy*, 101: 105201. https://doi.org/10.1016/j.landusepol.2020.105201.

Babaeian Jelodar, M. and Sutrisna, M. (2022). Guest editorial: Working smarter by applying advanced technologies in construction: Enhancing capacity and capability in construction sector for infrastructure project delivery. *Journal of Engineering, Design and Technology*, 20(4): 861–865. https://doi.org/10.1108/JEDT-08-2022-744/FULL/PDF.

Bilal, M., Iftikhar Ali, M., Rasheed, K. and Saad, S. (2023). Exploring adoption trends of knowledge management systems in the Software Industry's SMEs: A comprehensive investigation of enablers and barriers. *2023 4th International Conference on Data Analytics for Business and Industry (ICDABI)*, 598–604. https://doi.org/10.1109/ICDABI60145.2023.10629373.

Ding, L., Zhou, Y. and Akinci, B. (2014). Building Information Modeling (BIM) application framework: The process of expanding from 3D to computable nD. *Automation in Construction*, 46: 82–93. https://doi.org/10.1016/j.autcon.2014.04.009.

Erdogan, B., Abbott, C. and Aouad, G. (2010). Construction in year 2030: Developing an information technology vision. *Philosophical Transactions of the Royal Society A: Mathematical, Physical and Engineering Sciences*, 368(1924): 3551–3565. https://doi.org/10.1098/RSTA.2010.0076.

Gallala, A., Kumar, A. A., Hichri, B. and Plapper, P. (2022). Digital Twin for human–robot interactions by means of Industry 4.0 enabling technologies. *Sensors*, 22(13): 4950. https://doi.org/10.3390/S22134950.

Ghosh, S. K. and Ghosh, S. K. (2021). Circular economy: Recent trends in global perspective. *Circular Economy: Recent Trends in Global Perspective*, 1–573. https://doi.org/10.1007/978-981-16-0913-8/COVER.

Grübel, J., Thrash, T., Aguilar, L., Gath-Morad, M., Chatain, J. et al. (2022). The Hitchhiker's guide to fused twins: A review of access to Digital Twins *in situ* in smart cities. *Remote Sensing*, 14(13): 3095. https://doi.org/10.3390/RS14133095.

Haris, M., Saad, S., Ammad, S. and Rasheed, K. (2024). AI in fabrication and construction. In *AI in Material Science* (pp. 169–192). CRC Press. https://doi.org/10.1201/9781003438489-8.

Huang, Z., Shen, Y., Li, J., Fey, M., Brecher, C. et al. (2021). A survey on AI-driven Digital Twins in Industry 4.0: Smart manufacturing and advanced robotics. *Sensors*, 21(19): 6340. https://doi.org/10.3390/S21196340.

Javaid, M., Haleem, A. and Suman, R. (2023). Digital Twin applications toward Industry 4.0: A review. *Cognitive Robotics*, 3: 71–92. https://doi.org/10.1016/j.cogr.2023.04.003.

Khan, M. I., Khan, N., Saad, S. and Khan, M. W. (2024). Artificial Intelligence in construction materials. In *AI in Material Science* (pp. 47–64). CRC Press. https://doi.org/10.1201/9781003438489-3.

Kousi, N., Gkournelos, C., Aivaliotis, S., Giannoulis, C., Michalos, G. et al. (2019). Digital Twin for adaptation of robots' behavior in flexible robotic assembly lines. *Procedia Manufacturing*, 28: 121–126. https://doi.org/10.1016/J.PROMFG.2018.12.020.

Mazumder, A., Sahed, M. F., Tasneem, Z., Das, P., Badal, F. R. et al. (2023). Towards next generation digital twin in robotics: Trends, scopes, challenges, and future. *Heliyon*, 9(2): e13359. https://doi.org/10.1016/j.heliyon.2023.e13359.

Raja Santhi, A. and Muthuswamy, P. (2023). Industry 5.0 or industry 4.0S? Introduction to Industry 4.0 and a peek into the prospective Industry 5.0 technologies. *International Journal on Interactive Design and Manufacturing (IJIDeM)*, 17(2): 947–979. https://doi.org/10.1007/S12008-023-01217-8.

Rasheed, K., Saad, S. and Ammad, S. (2023). Green financing for sustainable economic growth: exploring critical factors towards resilient economies. *2023 International Conference on Sustainable Islamic Business and Finance (SIBF)*, 104–111. https://doi.org/10.1109/SIBF60067.2023.10379984.

Rasheed, K., Saad, S., Ammad, S. and Bashir, M. T. (2024). Industry 4.0 and construction. In *AI in Material Science* (pp. 65–87). CRC Press. https://doi.org/10.1201/9781003438489-4.

Rasheed, K., Waqas, M., Saad, S., Zaland, A., Ali, Z. et al. (2024). Ethical and societal implications of AI in construction. In *AI in Material Science* (pp. 216–238). CRC Press. https://doi.org/10.1201/9781003438489-10.

Rasheed, K., Zaland, A., Saad, S., Ammad, S., Rostami, A. et al. (2024). History of AI. In *AI in Material Science* (pp. 15–46). CRC Press. https://doi.org/10.1201/9781003438489-2.

Rezaei, Z., Vahidnia, M. H., Aghamohammadi, H., Azizi, Z., Behzadi, S. et al. (2023). Digital Twins and 3D information modeling in a smart city for traffic controlling: A review. *Journal of Geography and Cartography*, 6(1): 1865. https://doi.org/10.24294/JGC.V6I1.1865.

Saad, S., Haris, M., Ammad, S. and Rasheed, K. (2024). AI-assisted building design. In *AI in Material Science* (pp. 143–168). CRC Press. https://doi.org/10.1201/9781003438489-7.

Saad, S., Rasheed, K., Ammad, S., Ali, Z., Zaland, A. et al. (2024). The conclusive role of AI in Industry 4.0 construction materials. In *AI in Material Science* (pp. 262–270). CRC Press. https://doi.org/10.1201/9781003438489-12.

Saad, S., Rasheed, K., Ammad, S., Waqas Khan, M., Zaland, A. et al. (2024). Is AI the architect of tomorrow's materials in the age of Industry 4.0? In *AI in Material Science* (pp. 1–14). CRC Press. https://doi.org/10.1201/9781003438489-1.

Saad, S., Zaland, A., Ammad, S. and Rasheed, K. (2024). Robotics in construction. In *AI in Material Science* (pp. 115–142). CRC Press. https://doi.org/10.1201/9781003438489-6.

Segovia, M. and Garcia-Alfaro, J. (2022). Design, modeling and implementation of Digital Twins. *Sensors*, 22(14): 5396. https://doi.org/10.3390/S22145396.

Thacker, S., Adshead, D., Fay, M., Hallegatte, S., Harvey, M. et al. (2019). Infrastructure for sustainable development. *Nature Sustainability*, 2(4): 324–331. https://doi.org/10.1038/s41893-019-0256-8.

Vial, G. (2021). Understanding digital transformation: A review and a research agenda. *Managing Digital Transformation*, 13–66. https://doi.org/10.4324/9781003008637-4.

Vrana, J. and Singh, R. (2021). NDE 4.0—A design thinking perspective. *Journal of Nondestructive Evaluation*, 40(1): 1–24. https://doi.org/10.1007/S10921-020-00735-9/FIGURES/7.

Wang, H., Chen, X., Jia, F. and Cheng, X. (2023). Digital twin-supported smart city: Status, challenges and future research directions. *Expert Systems with Applications*, 217. https://doi.org/10.1016/J.ESWA.2023.119531.

Waqas Khan, M., Saad, S., Ammad, S., Rasheed, K., Jamal, Q. et al. (2024). Smart infrastructure and AI. In *AI in Material Science* (pp. 193–215). CRC Press. https://doi.org/10.1201/9781003438489-9.

Zaland, A., Saad, S., Rasheed, K. and Ammad, S. (2024). Automation in the construction industry. In *AI in Material Science* (pp. 88–114). CRC Press. https://doi.org/10.1201/9781003438489-5.

Index

D

E

M

N

O

P

Q

R

About the Editors

Dr. Syed Saad is an Engineering Lecturer at Sandwell College in the United Kingdom, with a Ph.D. from Universiti Teknologi Petronas (UTP), Malaysia and a Masters from Liverpool John Moores University, UK. He has received notable accolades, including a PMI Future 50 which recognizes top-tier professionals who demonstrate extraordinary potential to shape the future of project management across industries, outstanding research performance award from UTP and endorsements from the European Union and Honorary Consul General of the Republic of Tunisia in Pakistan. In his previous role as an Assistant Professor at CECOS University, he was awarded an outstanding research cluster head award. Dr. Saad has successfully helped over 1,500 students obtain free certifications from leading tech companies. With more than 70 published academic articles and a book on AI in Material Sciences, he serves as a professional reviewer for major journals and focuses his research on advanced civil engineering technologies such as Building Information Modeling (BIM), the Internet of Things (IoT), robotics, digital twins, and artificial intelligence. As Chief Research Officer, he has developed innovative research hierarchies and pioneered non-traditional teaching methods that emphasize student mental wellness. Additionally, he promotes professional networking through LinkedIn, empowering students to pursue international scholarships and publication opportunities while earning recognition as "LinkedIn Top Voices" and "Mendeley Research Advisors" in the region. Dr. Saad is also proficient in programming languages and tools including MATLAB, R-Studio, Python, C#, and VBA.

Kumeel Rasheed is a computer engineer by profession and has a M.Sc. degree in Project Management (Cum Laude distinction) from Bahria University, Pakistan. His research interest is in the domain of Circular Economy, Corporate Social Responsibility, Decisions Support Systems, Artificial Intelligence, Internet of Things, Industry 4.0/5.0 and Engineering Project Management. He has earned multiple certifications in IT and management from renowned leading corporations such as

Microsoft, Oracle, IBM, VMware, PMI, Cisco, Salesforce, International Scrum Institute. He is accredited as MCSE, Scrum Master Professional, ESG Professional, OCI Generative AI Certified Professional, Risk Management Expert, Salesforce Certified Associate, Design Thinking Practitioner, and Cyber Ops Associate. He held the position of Research Associate at CECOS University of IT and Emerging Sciences, Pakistan, where he was honoured with the Best Research Associate Award for the academic year. Kumeel has also received appreciation and endorsement from Honorary Consul General of the Republic of Tunisia, Peshawar and Dean of the Honorary Consular Corps (KP Chapter), Pakistan, for his co-edited book *'AI in Material Science: Revolutionizing Construction in the Age of Industry 4.0'*, that explores AI's impact on material science within Industry 4.0, focusing on innovation and optimization in construction and fabrication. He has delivered many workshops on practical research applications and presently occupy the position of Chief Research Officer in a research cluster. He has a passion for technology and innovation and is keen to advance sustainable and transformative practices in the fields of engineering and management.

Syed Ammad is a Ph.D. candidate in the School of Engineering and Built Environment at Griffith University, Gold Coast, Australia. He serves as a Graduate Research Assistant and Tutor at Griffith University, where he teaches Project Management Principles across the Gold Coast and Nathan campuses. His doctoral research focuses on mitigating fall-related hazards in construction, particularly during overhead work and rooftop activities. Ammad holds a master's degree in construction safety from Universiti Teknologi Petronas, Malaysia, where he also worked as a tutor and lab demonstrator. He previously served as a lecturer at CECOS University in Peshawar, Pakistan. His academic career includes publishing approximately 60 research papers as both first and co-author in prestigious journals, alongside serving as a peer reviewer for leading international publications. He co-edited the book AI in Material Science: Revolutionizing Construction in the Age of Industry 4.0, published in 2024, which explores the transformative impact of artificial intelligence on material science and construction processes within the industry 4.0 framework. His research interests encompass construction safety, fall prevention in construction, and the integration of artificial intelligence into construction management. Ammad's current work involves applying advanced technologies such as MATLAB, Python, and OpenPose for posture recognition and fall-related data collection, contributing innovative solutions to building information modeling and enhancing safety protocols within the construction industry.